FOCUS ON CELLULAR SIGNALLING RESEARCH

FOCUS ON CELLULAR SIGNALLING RESEARCH

DOROTHY T. LEEDS
EDITOR

Nova Science Publishers, Inc.
New York

For permission to use material from this book please contact us:
Telephone 631-231-7269; Fax 631-231-8175
Web Site: http://www.novapublishers.com

Library of Congress Cataloging-in-Publication Data
Available upon request.

ISBN 1-59454-619-3

Published by Nova Science Publishers, Inc. ✣ New York

Contents

Preface

Cell signaling is the communication inside a cell or with its environment. It plays an important role in cellular growth, survival and apoptosis. For instance, the growth of a cell requires production of certain proteins. The genes that encode these proteins are located in DNA inside the nucleus while the initial signal for cellular growth may come from outside the cell. To induce the expression of these genes, the external signal must pass through specific pathways to reach these genes in DNA. The signaling pathway involves several proteins. If any one of them becomes abnormal or non-functional, the cellular process would be altered, which may cause disease. Therefore, knowing the detailed pathway is very important in understanding the pathogenesis of diseases.

Angiotensin II is a biologically active peptide that acts both as a hemodynamic regulator and as a growth factor. The mitogenic effects of angiotensin II have been linked to pathophysiological conditions such as hypertension, atherosclerosis and restenosis. The cellular responses to angiotensin II is directed by specific G-protein coupled receptors. There are two major classes of angiotensin II receptors that have been pharmacologically defined and are now commonly referred to as the AT_1 and AT_2 receptor subtypes. The use of non-peptidic inhibitors for each of these receptor subtypes has demonstrated that the hemodynamic manifestations of angiotensin II are mediated by the AT_1 receptor. AT_1 receptors contain the structural features of the seven transmembrane, G-protein-coupled receptor superfamily and thus are structurally different from cytokine or growth factor receptors. Angiotensin II binding to the AT_1 receptor triggers the activation of intracellular signalling pathways. These signalling pathways are the key determinants for the angiotensin II-mediated cardiovascular diseases. Angiotensin II signalling is regulated positively and negatively through various mechanisms. The combined effects of the positive and negative regulation results in execution of signals of adequate strength for the appropriate time within precise spatial boundaries thereby avoiding potentially deleterious persistent signaling. The mechanisms involved in regulation of angiotensin II signalling are beginning to be emerged. However, recent data from several labs suggested that angiotensin II signalling is activated or inhibited by complex and highly regulated mechanisms that involve heterotrimeric G-proteins, tyrosine kinases, tyrosine phosphatases, scaffold/adaptor molecules, internalization/desensitization, and structural organization of the AT_1 receptor. Chapter I discusses some advances published over the last few years concerning the factors that regulate angiotensin II-mediated signalling. Given the critical roles of angiotensin II signalling in numerous fundamental cellular processes, it is necessary to study the role of these pathways in the pathophysiology of vascular disease.

Glucocorticoid receptor (GR) and aryl hydrocarbon receptor (AhR) are two entirely different individuals, the former being a typical steroid receptor integrated in endogenous affairs, the latter being the first described xeno-receptor, so-called dioxin receptor. Interestingly, these two receptors share several common features: 1) involvement in essential cellular processes such as proliferation, differentiation, inflammatory response, and gene expression; 2) essential role in drug metabolism, i.e. direct and/or indirect control of phase I, II, and III enzymes expression. 3) mechanism of their action in terms of ligand binding, nuclear import and export, and proteasome-mediated degradation, the last feature serving as negative feedback. 4) The dependence of their activity on the status of microtubules network. Despite voluminous data published, the role of microtubules in GR signaling is still controversial. Crucial appears the distinction between proliferating versus non-proliferating cells. While perturbation of microtubules in human cervix carcinoma cells (HeLa) resulted in rapid time- and dose-dependent, proteasome-mediated degradation of GR protein, microtubules disruption in primary human hepatocytes down-regulated GR driven genes by a mechanism involving inhibition of GR nuclear import. The mentioned similarities between GR and AhR also raise the question of the role of microtubules in AhR signaling. Nocodazole treatment caused inhibition of CYP1A1 dioxin-dependent inducibility in murine hepatoma cells (Hepa-1), but other data regarding microtubules integrity influence on AhR activity are much more scarce. Another common denominator for AhR and GR signaling is the ambiguity whether the alteration of either receptor transcriptional activity is due to cell cycle arrest or due to direct involvement of microtubules integrity in AhR and GR signal transduction. Chapter II is a summary of the evidence on the AhR and GR signaling under the microtubules network perturbation and formulates a hypothesis on the physiological role of both receptors, with regard to the implications in drug metabolism.

These results in chapter III suggest that radiotherapy blocks cellular responsiveness to BMP-2 induced osteoblastic differentiation in a time and dose dependent manner within a short window of time. The primary mechanism seems to be down-regulation of the BMP-2/receptor complex. Further analyses should differentiate between an influence on BMP-receptor structure, BMP-receptor function (binding capacity) or a combination. This effect is clearly different from the way ionizing irradiation acts in cytotoxic radiotherapy. Taking a possible down-regulation of the BMP-2/receptor complex in account, we proposed that the radiation of C2C12 cells influences the BMP receptor signalling by interfering with the formation or reshuffling of the BMP receptors on the cell surface and will cause changes in the localization of the BMP receptors in specific domains by down regulation of proteins specific for BMP regulation.

The biological function of proteins is determined by their cellular localization and subsequent interactions with other factors within a given subcellular compartment. Therefore, it is critical to understand how proteins move to and from the sites where they exert biological effects and how this mechanism of movement is regulated. This concept becomes particularly interesting when soluble signalling factors such as STATs, p53, NFkB, MAPKs, C/EBPs, steroid receptors or cyclins are involved. Soluble proteins are not confined to the cytoplasm or the nucleus in a static manner, but they shuttle dynamically between subcellular compartments regardless of where they are primarily localized under certain biological conditions. Consistent with this concept —and not surprisingly—protein mistargeting leads to a number of pathologies. Ideally, most of these pathologies could be attenuated or even prevented if we were able to regulate the subcellular localization of those mistargeted proteins; i.e. by interfering with the molecular machinery for protein movement and/or by regulating the function of anchoring factors of a given compartment. Nowadays, we know no more than the basics about regulatory mechanisms for protein anchoring, so we still have more questions and doubts than answers and certainties. On the other hand, the molecular mechanism by which soluble proteins move in the cell is an unsolved biological conundrum. In this regard, movement has always been assumed to occur in a stochastic manner by simple diffusion. This oversimplified model has been accepted as the driving mechanism for protein movement in both the cytoplasm and the nucleus. Although heuristic, this notion contradicts the concept of efficiency for protein targeting and, even more importantly, it collides against the concepts of cellular compartmentalization and specificity of action of signalling proteins. In chapter IV, the authors describe a novel model in which the cytoplasmic movement of some members of the nuclear receptor superfamily is regulated by their association with high molecular weight immunophilins. Based on their subnuclear distribution in different conditions, they also reformulate some classical concepts about the functional regulation of archetype hormone-regulated transcription factors such as steroid receptors and C/EBPs, and postulate the existence of a non-random mechanism for intranuclear trafficking of proteins.

The specificity and fidelity in cellular signal transduction is in part owed to the remarkable ability of signaling and adaptor proteins to exist in modular fashion. Such modular organization of proteins allows them to participate in a wide spectrum of signaling circuits simultaneously yet in a highly specific manner. This is the result of the specific interaction between protein modules found in a wide multitude of proteins and tailored to meet specific demands. Recent studies indicate that such protein modules may not always act independently in a one-on-one manner but rather their activity may be enhanced by the presence of a second copy in tandem. Such tandem repeats appear to act in a highly synergistic manner so that their biological function is greater than the sum of their parts. In chapter V, using bromodomains, SH2 domains, WW domains and PDZ domains as prototypes, the authors discuss the structure, function and mechanism of how such tandem repeats act in a co-operative manner and the biological implications of such an elegant design that nature seems to have adopted.

Calmodulin (CaM) is an essential protein in the model eukaryote *Dictyostelium discoideum* where it mediates numerous events including chemotaxis, gametogenesis, fertilization, and spore germination. Profiling using the CaM-binding overlay technique (CaMBOT) has revealed that well over four-dozen calcium-dependent and -independent CaM-binding proteins (CaMBPs) are present in *Dictyostelium*, some of which are linked to these CaM-mediated processes. Individual CaMBPs have been localized to specific sub-cellular locales and to show varying patterns of developmental expression. CaMBOT also is used to isolate cDNAs encoding CaMBPs from an expression library from mid-to-late multicellular development of *Dictyostelium*. Previously studied CaMBPs such as calcineurin A (CNA) and regulatory myosin light chain (RMLC) were identified in this way. Novel proteins were also identified. Nucleomorphin is a novel nuclear CaMBP, expressed as two main isoforms NumA1 and 2, which localizes to the nucleoplasmic periphery. NumA has at least one bipartite NLS (^{48}KKSYQDPEIIAHSRPRK^{66}) and a single CaM-binding domain (^{171}EDVSRFIKGKLLQKQQKIYKDLERF^{195}) that interacts with CaM in both a Ca^{2+}-dependent and Ca^{2+}-independent manner. NumA1 contains an extensive glu/asp or DEED repeat that regulates nuclear number. Yeast two hybrid and co-immunoprecipitation studies have identified calcium binding protein CBP4a as an interacting protein that binds to the DEED repeat. NumA2 also has a breast cancer C-terminus (BRCT) domain, two poly-N tracts plus a second, putative CaMBD. Proteins not previously known to be CaMBPs were also identified by CaMBOT probing of the *Dictyostelium* developmental expression library, including phosphoglycerate kinase, thymidine kinase and histone H1. *Dictyostelium* phosphoglycerate kinase (DdPGK) binds CaM in a Ca^{2+}-dependent manner with CaM negatively regulating its activity *in vitro*. The identified CaMBD shows 80% identity with PGKs from diverse organisms and is localized adjacent to mutation sites underlying certain human diseases. Thymidine kinase (DdTK1) is a Ca^{2+}-dependent CaMBP and its *in vitro* activity is slightly enhanced by CaM and inhibited by CaM antagonists. Histone H1 (DdH1) was isolated as a CaMBP and binds CaM *in vitro* but its mode of CaM-binding remains elusive. Each of the CaMBPs the authors have identified is discussed, in chapter VI, in terms of its individual roles as well as its potential interaction with other CaMBPs and other proteins in regulating specific CaM-dependent cellular processes including the cell cycle.

In: Focus on Cellular Signalling Research
Editor: Dorothy T. Leeds, pp. 1-27
ISBN 1-59454-619-3

Chapter I

Understanding AT_1 Receptor Signalling

Peter P. Sayeski[1] and M. Showkat Ali[2*]
[1] Department of Physiology and Functional Genomics,
University of Florida College of Medicine
[2]Department of Pathology and Laboratory Medicine,
Emory University School of Medicine

Abstract

Angiotensin II is a biologically active peptide that acts both as a hemodynamic regulator and as a growth factor. The mitogenic effects of angiotensin II have been linked to pathophysiological conditions such as hypertension, atherosclerosis and restenosis. The cellular responses to angiotensin II is directed by specific G-protein coupled receptors. There are two major classes of angiotensin II receptors that have been pharmacologically defined and are now commonly referred to as the AT_1 and AT_2 receptor subtypes. The use of non-peptidic inhibitors for each of these receptor subtypes has demonstrated that the hemodynamic manifestations of angiotensin II are mediated by the AT_1 receptor. AT_1 receptors contain the structural features of the seven transmembrane, G-protein-coupled receptor superfamily and thus are structurally different from cytokine or growth factor receptors. Angiotensin II binding to the AT_1 receptor triggers the activation of intracellular signalling pathways. These signalling pathways are the key determinants for the angiotensin II-mediated cardiovascular diseases. Angiotensin II signalling is regulated positively and negatively through various mechanisms. The combined effects of the positive and negative regulation results in execution of signals of adequate strength for the appropriate time within precise spatial boundaries thereby avoiding potentially deleterious persistent signaling. The mechanisms involved in regulation of angiotensin II signalling are beginning to be emerged. However, recent data from several labs suggested that angiotensin II signalling is activated or inhibited by complex and highly regulated mechanisms that involve heterotrimeric G-proteins, tyrosine kinases, tyrosine phosphatases, scaffold/adaptor molecules, internalization/desensitization, and structural

[*] Corresponding author: Department of Pathology and Laboratory Medicine, Emory University School of Medicine, 1639 Pierce Drive, 7107 WMB; Atlanta, GA 30322

organization of the AT_1 receptor. Here we discuss some advances published over the last few years concerning the factors that regulate angiotensin II-mediated signalling. Given the critical roles of angiotensin II signalling in numerous fundamental cellular processes, it is necessary to study the role of these pathways in the pathophysiology of vascular disease.

Introduction

One hypothesis consistent with current data is that, the binding of angiotensin II to the AT_1 receptor activates growth-promoting, intracellular signal transduction pathways which result in an orchestrated growth response of the cell. In some cases, however, when the growth response is not well coordinated, this may result in overt clinical complications. An important issue therefore is to understand the signal transduction mechanisms by which the AT_1 receptor converts ligand binding into a cellular growth response. The hope is that the delineation of these specific signalling pathways will result in a better understanding of the cellular growth responses that sometimes lead to vascular disease states.

With respect to growth factor signal transduction mechanisms, a well understood model pertains to epidermal growth factor (EGF) and platelet derived growth factor (PDGF). These ligands bind high affinity cell surface receptors. These receptors are characterized as having an amino terminal extracellular domain, a single membrane-spanning domain and an intracellular domain. The binding of ligand to the growth factor receptor triggers the activation of an intrinsic tyrosine kinase domain, located on the intracellular portion of the receptor. The activated receptor then autophosphorylates its own tyrosine residues that are contained on the intracellular domain of the receptor. These phosphotyrosine moieties serve as the site for the assembly of a signalling complex of intracellular molecules that are catalytically active. For example, activated growth factor receptors bind Src family tyrosine kinases (Pawson, 1997). The binding of a Src family kinase to a receptor phosphotyrosine induces a conformational change in the protein which leads to enzyme activation and the subsequent generation of a tyrosine phosphorylation signalling cascade. These signaling cascades terminate with the activation of growth promoting genes within the nucleus. Thus, in a mechanistic fashion, the binding of EGF or PDGF to its respective receptor, initiates the process by which changes in the external milieu are detected at the cell surface and transcriptional responses are adjusted accordingly within the nucleus.

The AT_1 receptor was cloned in 1991 and shown to belong to the seven transmembrane superfamily of receptors (Murphy et al., 1991; Sasaki et al., 1991). This class of receptors is different from growth factor receptors in that they contain an extracellular amino terminus, discreet hydrophobic domains that transverse the cell membrane seven times and end with an intracellular carboxy terminal tail. The signalling mechanism of seven transmembrane receptors has traditionally been portrayed as mechanistically different from that of growth factor receptors; the signalling mechanisms of these receptors is dependent on the dissociation of heterotrimeric GTP-binding proteins, or simply, heterotrimeric G proteins. Ligand binding induces the dissociation of the $G_{\alpha\beta\gamma}$ heterotrimeric complex into the G_α and $G_{\beta\gamma}$ subunits. Each subunit acts as a signal transducer to activate effector molecules such as adenylyl cyclase, phospholipase C and ion channels. Thus, it is thought that seven transmembrane surface receptors mediate the flow of information from the outside of the cell to the inside of the cell via heterotrimeric G proteins and not tyrosine phosphorylation signaling cascades. As is shown below, this concept of signaling by seven transmembrane surface receptors is not correct. Recent work by numerous independent laboratories has demonstrated that in addition to the AT_1 receptor, the seven transmembrane surface receptors that bind vasopressin, bombesin, bradykinin, endothelin, platelet activating factor, lysophosphatidic acid, α-melanocyte stimulating hormone, isoproterenol, phenylephrine and norepinephrine also elicit tyrosine phosphorylation signaling cascades (Buggy, 1998; Dhar and Shukla, 1994; Erpel and Courtneidge, 1995; Luttrell et al., 1999; Peeler et al., 1996; Sasaguri et al., 2000; Zhong et al., 2000). Thus, the paradigm of a receptor based tyrosine kinase signaling applies not only to growth factor receptors, but also to seven transmembrane surface receptors as well.

I. G-Protein-Dependent Signaling Pathways

Heterotrimeric G proteins mediate the transduction of signals from the outside to the inside of the cell by coupling ligand occupancy of GPCRs with the activation of intracellular effector molecules. Effector molecules that are activated by G proteins include adenylyl cyclase, phospholipase C, and ion channels such as K^+ and Ca^{2+} channels.

Heterotrimeric G proteins are guanine nucleotide binding proteins. Because of their ability to hydrolyze GTP, they belong to the GTPase superfamily of proteins. This superfamily is subdivided into two families, namely the monomeric GTPases and heterotrimeric G proteins. The monomeric GTPases are cytoplasmic proteins that become membrane associated upon activation and are involved in a diverse set of signal transduction events. Members include the Rho, Ras, and Arf proteins. In contrast, the heterotrimeric G proteins are *membrane bound* molecules which appear to couple exclusively with seven transmembrane spanning receptors. Heterotrimeric G proteins are formed by three distinct protein chains designated as the α-, β-, and γ- subunits. These protein subunits are distinguishable by several criteria, including size: α- range in size from 39 to 52 kDaltons, β- are 36 kDaltons, and γ- are 6 to 8 kDaltons. To date, twenty three different α-, six β-, and eleven γ- subunits have been identified. The α- subunits have been further subdivided into four families designated $G_{\alpha s}$, $G_{\alpha i}$, $G_{\alpha q}$, and $G_{\alpha 12}$. The expression of these different subunits in a cell could result in numerous $G_{\alpha\beta\gamma}$ combinations assuming that all are expressed equally and all associate randomly. However, it appears that 1) each cell expresses different subunits, 2) the association of these subunits may not be random, and 3) the interaction of a specific $G_{\alpha\beta\gamma}$ trimer with a specific receptor is unique. This last point has been observed in several instances. For example, in the GH_3 pituitary cell line, the inhibition of Ca^{2+} channel activity by somatostatin is mediated by the $\alpha_{i\text{-}o1}/\beta_1/\gamma_3$ complex whereas inhibition of the same Ca^{2+} channel by the M4 muscarinic receptor occurs through the $\alpha_{i\text{-}o1}/\beta_3/\gamma_4$-trimer (Barnard, 1992). Thus, one level of specificity for a given GPCR signal is determined by the expression of the appropriate receptor, the expression of the appropriate $G_{\alpha\beta\gamma}$ subunits, and the specific interaction of these subunits with the receptor.

In the inactive state, the $G_{\alpha\beta\gamma}$ trimer is bound to GDP via the Gα- subunit. Activation of heterotrimeric G proteins is initiated by its interaction with specific cytoplasmic segments of a ligand-activated GPCR. As a result, GDP is released from the Gα- subunit. The Gα-subunit then binds GTP and Mg^{2+} with very high affinity and dissociates from the Gβγ complex. Both the activated Gα-GTP and the Gβγ subunits modulate effector proteins until hydrolysis of the GTP bound to the Gα subunit terminates signaling via the intrinsic GTPase activity contained within the α- subunit. Following GTP hydrolysis, the inactive Gα-GDP subunit dissociates from the effector molecule and re-associates with the Gβγ subunit to return to the inactive $G_{\alpha\beta\gamma}$ heterotrimer.

The primary amino acid structure of the AT_1 receptor indicated that it fell within the GPCR superfamily. That said, efforts were made to establish whether it was in fact capable of G protein activation. The most reliable way to demonstrate this was to exploit two characteristics that are hallmarks of G protein activation: its membrane association and its ability, when activated, to bind GTP with very high affinity. Plasma membranes can be prepared from angiotensin II treated cells and incubated with a non-hydrolyzable analogue of GTP, ^{35}S-GTPγS. The activated Gα- subunit is capable of binding GTPγS with very high affinity, but the γ-sulfur makes it insensitive to the GTPase activity of the α- subunit. Heterotrimeric G protein activation is directly quantitated by the amount of radioactivity bound to the membranes. Early studies indicated that angiotensin II stimulated-membranes were in fact capable of activating heterotrimeric G proteins as measured by the amount of ^{35}S-GTPγS bound to the filters when compared with unstimulated membrane controls (Crawford et al., 1992; Inagami et al., 1992). These results set in motion a new series of studies to identify the specific complement of G protein subunits that are coupled to AT_1 receptor activation.

The addition of purified Gα subunits into these *in vitro* assays demonstrated that the AT_1 receptor can couple to $G_{\alpha q}$, $G_{\alpha q/11}$, and $G_{i/0}$ (Kai et al., 1996). However, these studies had several limitations: the measurements were done outside the cell and the appropriate complement of Gβγ subunits could not be identified. In subsequent work, a functional assay was set up whereby the angiotensin II-dependent entry of extracellular Ca^{2+} via the L-type Ca^{2+} channel in cultured rat portal vein myocytes was assessed (Macrez-Lepretre et al., 1997). Antisense oligonucleotides directed against various α-, β-, and γ- subunits were injected into cells and ligand-dependent Ca^{2+} entry was measured. The authors found that oligonucleotides directed against the mRNA coding for the $\alpha_{13}/\beta_1/\gamma_3$ subunits blocked the angiotensin II-mediated entry of extracellular Ca^{2+}. A corresponding reduction in $\alpha_{13}/\beta_1/\gamma_3$ protein levels was confirmed by the immunohistochemical staining of antisense injected cells. Furthermore, these studies suggested that the interaction of the heterotrimeric complex with the AT_1 receptor was mediated by the C terminal region of the $G_{\alpha 13}$ subunit, as addition of a synthetic peptide corresponding to the carboxyl terminal end of $G_{\alpha 13}$ inhibited, in a concentration-dependent manner, the angiotensin II-dependent Ca^{2+} response. Collectively, these studies demonstrated that activation of the AT_1 receptor by angiotensin II resulted in coupling to heterotrimeric G proteins. More importantly, the coupling to a specific heterotrimer G protein complex distal to AT_1 receptor activation was found to have a functional significance.

Another important signal transduction pathway in vascular cells that is distal to G protein activation is the protein kinase C (PKC) signaling pathway. PKCs are serine/threonine kinases and their substrates include proteins that are important in cellular proliferation such as histones and the myristolated alanine-rich C kinase substrates (MARCKS). There are currently eleven members belonging to the PKC family and they are differentiated by their mechanisms of activation (ie Ca^{2+}, phospholipids, diacylglycerol, or phorbol esters). The activation of phospholipases through both G protein-dependent and G protein-independent mechanisms causes PIP_2 to be hydrolyzed to IP_3 and diacylglycerol (DAG). DAG in turn activates PKC. The role of PKC in angiotensin II signaling has been studied by down regulating PKC activity through prolonged exposure to phorbol esters. Down regulation of PKC significantly reduced the angiotensin II-mediated induction of the early response genes and cellular proliferation of vascular smooth muscle cells (VSMC) (Leszczynski et al., 1996; Taubman et al., 1989). Thus, it is likely that events such as smooth muscle cell proliferation seen in some vascular diseases will have a PKC-dependent component.

Activation of G proteins by the AT_1 receptor initiated a search for the region of the AT_1 receptor that mediates G protein activation. Initial studies were done by generating CHO cell lines that stably expressed either the wild type AT_1 receptor or various AT_1 receptor mutants. These studies indicated that Asp^{74} was important in angiotensin II signaling as cell lines containing a point mutation at this residue were unable to produce an angiotensin II-dependent increase in inositol phosphates or an increase of intracellular Ca^{2+} (Bihoreau et al., 1993). Interestingly, it appears that the importance of Asp^{74} lies in the binding of angiotensin II and not the direct coupling to heterotrimeric G proteins. Subsequent studies have demonstrated that, in response to ligand, there is a rotation in the membrane helices that results in Asp^{74} forming a hydrogen bond with Tyr^{292} and that this appears to be critical for receptor activation (Inoue et al., 1997). Thus, the Asp^{74} mutant is unable to couple with G proteins because of altered binding affinities for ligand as opposed to disruption of the region that mediates G protein activation.

Subsequent experiments using the *in vitro* G protein activation assay described above indicated that two regions within the AT_1 receptor were important for the activation of $G_{\alpha\text{-}i1}$, $G_{\alpha\text{-}i2}$, and $G_{\alpha\text{-}o}$. The first region was AT_1 receptor amino acids 216-230 found within the third intracellular loop while the second region spanned amino acids 306-320 found within the intracellular carboxyl tail of the receptor (Shirai et al., 1995). Intracellular analysis quickly confirmed the importance of these two regions. Work by Escobedo demonstrated that amino acids 219-225, contained within the third intracellular loop of the receptor, were critical for coupling to $G_{\alpha\text{-}q}$ (Wang et al., 1995). In an elegant experiment, they conferred AT_1 receptor signaling properties onto the AT_2 receptor by substituting the third intracellular loops between the two receptors. In similar work that was published two years later, the region contained within the carboxyl terminal tail which mediates $G_{\alpha\text{-}q}$ activation was mapped to amino acids 312-314 (Sano et al., 1997). Finally, Kai used synthetic peptides to interfere with angiotensin II induced membrane GTPase activity. This approach suggested that the second intracellular loop (residues 125-137), the N-terminal region of the third intracellular loop (217-227) and a portion of the C-terminal tail (304-316) were important for this G-protein dependent process (Kai et al., 1998).

Collectively, these studies demonstrate that the AT_1 receptor couples with heterotrimeric G proteins and this coupling is transduced to effector molecules such as Ca^{2+} channels, phospholipase C, or adenylyl cyclase. Thus, G proteins play an integral part in angiotensin II signal transduction and it is very likely that they partake in the diversity of responses that follows AT_1 receptor activation.

II: Structural Requirement of the AT_1 Receptor Involved in the Regulation of Angiotensin II Signalling

The structure of the AT_1 receptor was deduced from the well known crystal structure of other G-protein coupled receptors (GPCR) such as the rhodopsin receptor. Since its discovery, many groups have used pharmacological and molecular biology approaches to elucidate various structure-function relationships of the AT_1 receptor. Following a computer based molecular modeling approach, it was proposed that like other G-protein coupled receptors, the AT_1 receptor undergoes a conformational change between its inactive and active states (de Gasparo et al., 2000). In the absence of angiotensin II binding, structural constraint keeps the receptor in an inactive state. Two lines of evidence support this hypothesis. First, replacement of amino acid Asn^{111} with glycine confers constitutive activity to the receptor (Groblewski et al., 1997). Second, mutation of Asp ^{74} in the second trans-membrane domain (Bihoreau et al., 1993) and Tyr^{292} (Marie et al., 1994) in the seventh transmembrane domain significantly inhibit IP3 production, a commonly used index to measure the receptor activation. It was hypothesized that these amino acids participate in the angiotensin II induced conformational change of the AT_1 receptor to the activated state (Thomas et al., 2000). Thus, it appears that intramolecular interactions might be of fundamental importance for the process of the AT_1 receptor activation.

Mutagenesis studies of the AT_1 receptor have identified specific amino acids that are essential for ligand binding, internalization, G-protein coupling and signal transduction. Briefly, whereas seven transmembrane helices are required for the proper folding and processing of the receptor, the amino terminal extracellular domain is required for ligand binding (Miura et al., 2003). The 3rd intracellular loop of the AT_1 receptor is well known in receptor activation and transducing various signalling pathways. Hines and colleagues (Hines et al., 2003) reported that mutation of amino acids 221 and 222 within the 3rd intracellular domain of the AT_1 receptor diminished the receptors ability to activate PLC β with little or no effect on the activation of p44/p42 MAP kinase. The change of the diacidic motif (Asp^{236}-Asp^{237}) in the 3rd intracellular loop of the AT_1 receptor did not impair the activation of PLC β and MAP kinase activation, but diminished GRK-mediated receptor phosphorylation (Olivares-Reyes et al., 2001). Seta and co-workers (Seta et al., 2002) reported that the AT_1 mutant receptor which lacked the DRY motif (Asp^{125}-Arg^{126}-Tyr^{127}) activated the MAP kinase pathway but failed to phosphorylate nuclear targets due to lack of inhibition of Crm-1-induced nuclear export of ERK. These results suggest that modification of individual amino acids provide a fine tuning system to functionally separate the angiotensin II-induced signalling pathways.

As indicated earlier, the classical signalling pathway of the AT_1 receptor is activation of heterotrimeric G-proteins. Previously it was reported that amino acids 216-230 found within the 3rd intracellular loop (Wang et al., 1995) and amino acids 306-320 found within the intracellular carboxy terminus (Sano et al., 1997) are important for the activation of G-proteins. In addition, the DRY motif (Asp^{125}-Arg^{126}-Tyr^{127}), a highly conserved triplet amino acid sequence located at the 2nd intracellular loop of the AT_1 receptor which is found in most G-protein coupled receptors couples to $G\alpha q$ as well as the G_i subunit of G-proteins (Shibata et al., 1996). Ohyma and colleagues (Ohyama et al., 2002) reported that Arg^{126} in the DRY motif is essential for coupling $G\alpha_q$, but not other heterotrimeric G-proteins. This hypothesis was further supported by the observations for $G\alpha_q$ coupled to GnRH (Arora et al., 1997), m1 muscarnic (Zhu et al., 1994) and N-fromyl peptide (Bennett et al., 2000) receptors. Thus, what is clear is that there exists no simple sequence determinant that can be solely attributed to G-protein coupling.

A structural feature that distinguishes the AT_1 receptor is the long intracellular carboxy terminus, a region of about 60 amino acids. From multiple studies, it has become apparent that the AT_1 receptor carboxy terminus tail containing potential tyrosine and serine phosphorylation sites plays a significant role in receptor activation, internalization, interaction with other proteins and emanating signaling events from the cell surface to the nucleus. For example, Hunyady and co-workers (Hunyady et al., 1994) reported that the STL motif (Ser^{335}-Thr^{336}-Leu^{337}) is responsible for receptor endocytosis. It should be noted that the carboxyl terminus exhibits several consensus sequences for PKC-mediated serine phosphorylation and has been demonstrated to be phosphorylated *ex vivo* (Kai et al., 1994; Paxton et al., 1994). Although there is a debate regarding tyrosine phosphorylation of the AT_1 receptor, work by Seta (Seta and Sadoshima, 2003) demonstrated that Tyr^{319} in the carboxyl terminus of the receptor is phosphorylated in response to angiotensin II. These authors also demonstrated the specific involvement of Tyr^{319} in the transactivation of the epidermal growth factor receptor (EGF-R) and cell proliferation, possibly through its interaction with SHP-2 and/or the EGFR.

The paradigm of STAT activation by a cytokine receptor is that STATs are transiently recruited to the receptor complex via binding to specific phosphotyrosine residues through their SH2 domains (Leonard and O'Shea, 1998). Receptor-bound STAT proteins are then tyrosine phosphorylated by Jak kinases associated with the membrane proximal portion of the receptor. It is well established that, similar to cytokine receptors, activation of the AT_1 receptor by angiotensin II promotes 1) STAT tyrosine phosphorylation, 2) STAT nuclear translocation, 3) STAT-DNA binding activity, and 4) STAT-dependent transcriptional activation (Bhat et al., 1995; Marrero et al., 1995). However, the biochemical sequence of events leading to STAT activation for the AT_1 receptor has not been defined. To investigate the role of AT_1 receptor tyrosine residues in STAT1 activation, we studied a mutant AT_1 receptor, termed M6, which contained phenylalanine substitutions for the 6 tyrosine residues naturally found in the 3rd intracellular loop (1 tyrosine) and the carboxy terminus (5 tyrosines). The ability of the mutant receptor to stimulate STAT1 tyrosine phosphorylation was studied in transiently transfected COS-7 cells and in stably transfected CHO cells. Both systems yielded identical results: in response to angiotensin II, the M6 AT_1 mutant receptor triggered STAT1 tyrosine phosphorylation equivalent to that observed with the wild type AT_1

receptor (Ali et al., 2000). These data suggest that although STAT1 phosphorylation is regulated by the AT_1 receptor, it is independent of these receptor tyrosine residues. Thus, STAT1 activation by the AT_1 receptor is biochemically different from the paradigm of cytokine receptors where receptor phosphotyrosine residues are critical for subsequent STAT activation.

The last ten years of research in the field has revealed that the AT_1 receptor interacts with numerous signaling proteins. Thus, it is apparent that the receptor can act as a scaffold protein. The first report to show that the AT_1 receptor acts as a scaffold was shown with the tyrosine kinase Jak2 (Marrero et al., 1995). This initial work was further corroborated by a subsequent finding that the ^{319}YIPP motif (Tyr^{319}-Ile^{320}-Pro^{321}-Pro^{322}) in the carboxy terminus of the AT_1 receptor is involved in the binding of Jak2 to the AT_1 receptor (Ali et al., 1997a). An unexpected twist in this avenue of investigation came when the Marrero group reported that SHP-2 (Marrero et al., 1998) and PLC $\gamma 1$ (Venema et al., 1998) bound the same YIPP motif on the AT_1 receptor. Tyr^{319} in the YIPP motif has also been implicated in binding EGFR in a phosphorylation dependent manner (Seta and Sadoshima, 2003). The manner by which the YIPP motif binds these proteins and contributes to their functional roles within the cell warrant further investigation. Of particular interest, Inada and coworkers reported that two prolines in the YIPP motif are important for angiotensin II binding to the AT_1 receptor and ligand induced transient increase of intracellular calcium (Inada et al., 2003). The implication of this study is that some intermediate mechanism must serve to tie together ligand binding to the AT_1 receptor and AT_1 receptor based signaling complexes. Whether the intermediate mechanism is a result of the disassociation of heterotrimeric G-proteins, a change of intracellular calcium, or some other mechanism remains to be determined. Although the amino acid(s) have not yet been mapped, an *in vitro* interaction between endothelial nitric oxide synthase (eNOS) and the carboxy terminal tail of the AT_1 receptor has been reported (Ju et al., 1998). This observation opens a new avenue to study the functional cross talk between the AT_1 receptor and eNOS. Subsequently, a yeast-two hybrid screening approach was employed using the carboxyl terminal tail (aa 295-359) as a bait to identify the trans membrane protein named, AT_1-receptor associated protein (ATRAP) (Daviet et al., 1999). Subsequently, it was demonstrated ATRAP is involved in the regulation of angiotensin II-induced transcriptional activity of the early response gene, *c-fos,* and thereby regulates cell proliferation (Lopez-Ilasaca et al., 2003). The Inagami group (Guo et al., 2003) following a similar yeast-two hybrid screening approach, isolated a novel protein termed ARAP1 (angiotensin II type 1 receptor associated protein). Following deletion mutation studies, it was concluded that amino acid residues 319-359 of the AT_1 receptor were binding ARAP1. While ATRAP regulates AT_1 receptor internalization and desensitization (Daviet et al., 1999), ARAP1 regulates cell surface expression of the receptor (Guo et al., 2003). Collectively, these results suggest that the paradigm of receptor based signaling complexes applies to seven transmembrane receptors as well as growth factor and cytokine receptors. Finally, the identification of the AT_1 receptor as a signaling nexus is an emerging field of study and might explain the pleiotropic effects of angiotensin II. Such a mechanism may be a key determinant for the regulation of angioetnsin II-mediated mitogenesis and subsequent growth-associated pathophysiology.

A long standing hypothesis for the G-protein coupled receptor was "one ligand one receptor". In recent years, work from several laboratories has radically changed this concept. Combined biophysical, mutagenesis and pharmacological studies suggest that GPCRs can exist as either dimers or oligomers. The oligomerization of G-protein coupled receptors in turn has been shown to play a key role in ligand binding, receptor trafficking, and signal transduction (Angers et al., 2002). The AT_1 receptor has been reported to homo- or heterodimerize in a ligand dependent manner. The stable heterodimer formation between the AT_1 receptor and the bradykynin receptor causes increased activation of G-proteins associated with changes in the intracellular trafficking of these receptors (AbdAlla et al., 2000). A heterodimer formation between AT_1 and AT_2 receptors inferred inhibition of AT_1 receptor-dependent signaling (AbdAlla et al., 2001). An elegant study demonstrated that oligomerization of wild type and a non-functional mutant of the AT_1 receptor exerted a dominant negative effect on $G\alpha q$ coupling, whereas ERK activation and β-arrestin recruitment remain unaffected (Hansen et al., 2004). These results suggest that the AT_1 receptor oligomerization (homo or heterodimer) has differential effects on different signaling systems. Finally, AT_1 receptor oligomerization has clearly emerged as an important functional signaling modality. It may be worthwhile to examine whether ligand-dependent oligomerization of the AT_1 receptor might contribute to the regulation of angiotensin II signaling. This investigative effort would be extremely challenging were it not for the fact that other investigators have succeeded in creating mutants (M3 muscaranic and CCR2 receptors) that fail to oligomerize *in vivo,* in response to ligand binding (Hernanz-Falcon et al., 2004; Zeng and Wess, 1999). Thus, oligomerization appears to be a universal phenomenon that provides an additional mechanism of modulation of AT_1 receptor functions and cross-talk with other GPCRs.

III. Termination and Transduction of Angiotensin II-Induced Activation of Map Kinase Pathway

The MAP kinase signaling pathway is well characterized and is activated by a variety of factors including growth factors, cytokines, hormones, and stress factors. The MAP kinase family consists of MAPK, MAPK activators (MEK, MKK or MAPKK) and a MEK activator (MAPKKK or MEKKK). In mammals, five distinguishable MAP kinase proteins have been identified. These include the ERK1/ERK2, ERK3/ERK4, ERK5, c-jun N-terminal kinase (JNK1/JNK2) and p38 MAPK. Interestingly, each downstream kinase serves as a substrate for an upstream activator (Schaeffer and Weber, 1999). In a receptor activation model, association of Shc with an activated receptor enables the receptor or associated kinase to phosphorylate Tyr^{317}, the Grb2 binding site on Shc (Sasaoka et al., 1994). This modification allows the Shc-Grb2-mSos complex to assemble at the plasma membrane in close proximity to Ras. Activated Ras causes Raf-1 hyperphosphorylation, an event associated with Raf-1 activation. Raf-1 in turn activates MAPKK and finally MAP kinase. G-protein coupled receptors, such as thrombin, endothelin, α-adrenergic receptor, and the AT_1 receptor have been reported to activate the MAP kinase signaling pathway. It was hypothesized that Gq coupled receptors generally utilize a Ras independent pathway involving protein kinase C, whereas Gi coupled receptors initiate a Ras dependent pathway (van Biesen et al., 1996). Previously published results indicate that MAP kinase activation is important for the expression of early response genes in response to angiotensin II (Gille et al., 1995). Furthermore, numerous studies have reported the role of ERK in regulating cardiac hypertrophy. For example, transgenic over-expression of the ERK inhibitor, MAP kinase phosphtase-1, limits hypertrophic growth (Bueno et al., 2001), whereas over-expression of the ERK activator, MAPK/ERK-1, promotes hypertrophic growth (Bueno et al., 2000). Because the MAP kinase pathway is critical for the induction of many genes crucial for vascular hypertrophy and hyperplasia, understanding the regulation of this pathway may provide novel insight into certain pathophysiological conditions such as hypertension and cardiac hypertrophy.

Angiotensin II induces the rapid tyrosine phosphorylation of Shc in VSMC, cardiac myocytes, and cardiac fibroblasts (Linseman et al., 1995; Sadoshima and Izumo, 1996; Schorb et al., 1994). Phosphorylated Shc serves as a docking protein for Grb2, thus linking the receptor to the Grb2-Sos complex and ultimately activating Ras. It was reported that Fyn (a Src family tyrosine kinase) is the Shc tyrosine kinase in cardiac myocytes (Sadoshima and Izumo, 1996). Work by Schieffer and colleagues found that MAP kinase stimulation by angiotensin II involves the sequential activation of c-Src, Ras activation, and Ras-Raf complex formation (Schieffer et al., 1996). At present, the detailed mechanism of MAP kinase activation by the G-protein coupled AT_1 receptor is not clear. In particular, the role of PKC in MAP kinase activation is still not completely understood. PKC depletion has no effect on MAP kinase activation in VSMC and bovine adrenal glomerulosa cells (Eguchi et al., 1996; Tian et al., 1998). In contrast, PKC depletion significantly reduces MAP kinase activation in AT_1 expressing CHO-K1 cells (Arai and Escobedo, 1996). There is also data suggesting that the activation of the Raf-1/ERK cascade by angiotensin II in kidney cells and cardiac myocytes requires PKC (Terada et al., 1995; Zou et al., 1996). Earlier studies suggested that angiotensin II-induced activation of ERK occurs via both EGF receptor transactivation-dependent and -independent pathways. The precise mechanism of the EGF receptor on ERK activation is unclear, but it was recently reported that angiotensin II-dependent EGFR transactivation in specific cell types is related to the differential involvement of matrix metalloproteinases-dependent heparin binding-EGF release (Shah et al., 2004). Collectively, these findings suggest that the mechanism of MAP kinase pathway activation by the AT_1 receptor differs among cell types.

The discovery of β-arrestins has highlighted an important central mechanism for the regulation of the MAP kinase signaling pathway. β-arrestin binding to the phosphorylated AT_1 receptor uncouple the receptor from heterotrimeric G-proteins and target the receptor to clathrin coated pits for endocytosis and ultimately termination of intracellular signals (Claing et al., 2002). Although β-arrestins were discovered in the context of G-protein coupled receptor desensitization, recently it has been demonstrated that they also initiate signals from the receptors that they 'desensitize'. New aspects of arrestin function include their capacity to act as scaffolds that activate the MAP kinase pathway (Luttrell et al., 1999). The Lefkowitz group demonstrated that angiotensin II dependent activation of ERK is differentially regulated by β-arrestins and G-proteins (Wei et al., 2004). The authors demonstrated that angiotensin II-induced activation of the MAP kinase pathway proceeds through two signaling pathways: heterotrimeric G-proteins and β-arrestins. Concurrent with this observation, the same group also reported that β-arrestin-dependent, angiotensin II-induced activation of ERK is enhanced by overexpression of G-protein coupled receptor kinase (GRK) 5 and 6, and reciprocally diminished by GRK2 and 3 (Kim et al., 2005). These results indicate that β-arrestins recruited in response to AT_1 receptor phosphorylation by different GRKs have distinct functional outcomes. Collectively, these results suggest that activation of the MAP kinase pathway by β-arrestins presents a new perspective in understanding the underlying mechanism of angiotensin II-induced activation of MAP kinase pathway.

Studies over the past few years have demonstrated that linker or adaptor proteins are important for the appropriate subcellular localization and amplification of the MAP kinase signaling pathway. This idea is reinforced by the findings that the GPCR kinase-interacting protein, GIT1, is required for sustained activation of ERK in response to angiotensin II (Yin et al., 2004). GIT1 links the AT_1 receptor and the EGFR to ERK1/2 activation by directly associating with MEK1. This protein:protein co-association is mediated by coiled-coil domains and a Spa2 homology domain of GIT1. Thus, GIT1 serves as a scaffold protein to facilitate c-Src-dependent activation of MEK1 and ERK1/2 in response to both GPCR and tyrosine kinase receptors. Shc is another adaptor/docking protein that acts as an amplifier of tyrosine kinase-mediated signaling pathways leading to ERK activation. The Shc family contains three related proteins termed p46, p52, and p66, which differ only in amino terminal sequence. A great deal of evidence has demonstrated that Shc plays critical regulatory roles in numerous cellular events. For example, deletion of the p66 allele, extends lifespan by approximately 30% in mice and confers resistance to oxidative stress (Migliaccio et al., 1999). Mice lacking Shc expression specifically in the T-cell lineage, together with inducible expression of dominant negative Shc in transgenic mice, revealed an essential and nonredundant role for Shc in thymic T-cell development (Zhang et al., 2002). Lai and Pawson (Lai and Pawson, 2000) provided the best evidence to date for the biological role of Shc *in vivo*. When exons 2 and 3 of ShcA (which encode the PTB domain) were deleted by specific gene targeting ($ShcA^{\Delta ex2/3}$), the homozygous mutant is embryonic lethal with severe defects in heart and blood vessel development. This study demonstrates the importance of examining the physiological functions of Shc in various cardiovascular diseases. Since Shc phosphorylation is necessary to trigger MAP kinase activation, tyrosine phosphorylation of Shc was measured in VSMC. Of the three Shc isoforms, p66 and p52 were preferentially phosphorylated in response to angiotensin II. Shc is reported to be phosphorylated by several non-receptor kinases including c-Src, Jak2, and PI3 kinase (He et al., 1995; McGlade et al., 1992; Sato et al., 1998). To identify the specific kinase responsible for angiotensin II-induced phosphorylation of Shc in VSMC, cells were treated with AG-490, PP2, or Wortmannin (Burgering and Coffer, 1995; Hanke et al., 1996; Meydan et al., 1996). These are specific pharmacological inhibitors of Jak2, Src, and P-I-3 kinase, respectively. Only PP2, the Src family kinase specific inhibitor, effectively inhibited Shc phosphorylation in VSMC. To demonstrate that c-Src specifically phosphorylates tyrosine residues on Shc, VSMC were grown and transduced with either an adenovirus vector encoding kinase inactive Src (KI-Src) or a control adenovirus encoding LacZ (Ad. LacZ). Angiotensin II-induced Shc tyrosine phosphorylation was markedly inhibited in VSMC infected with kinase inactive Src when compared to LacZ controls. In summary, these results implicate c-Src as the tyrosine kinase that phosphorylates Shc in VSMC (Sayeski and Ali, 2003).

Angiotensin II is a potent mitogen in various cell types. Here, we provide evidence demonstrating the critical role of the Shc/Grb2/ERK2 pathway in mediating angiotensin II-dependent proliferative effects (Sayeski and Ali, 2003). Specifically, its effect on VSMC proliferation was examined by measuring the incorporation of [^{3}H] thymidine, an indicator of DNA replication. VSMC were serum starved for 48 h followed by treatment with angiotensin II for 18 h. [^{3}H] thymidine was then added to the medium and thymidine incorporation was measured 6 h later. We found that treating VSMC with 100 nM angiotensin II increased [^{3}H]-thymidine incorporation, with maximal increase (3-4 fold) occurring 18 h after ligand addition. To study the role of the MAP kinase pathway in angiotensin II-induced cell growth, VSMC were pretreated with either DMSO, the c-Src inhibitor PP2, or the MEK1/2 inhibitor PD98059. The cells were then left untreated or treated with 100 nM angiotensin II for 18 h. [^{3}H]-thymidine incorporation was then measured from all groups. Angiotensin II treatment resulted in a statistically significant, four-fold increase in DNA synthesis with respect to untreated control cells. Pretreatment with either PP2 (20 μM) or PD98059 (100 μM) eliminated the statistically significant angiotensin II proliferative growth response. Several laboratories have reported that the expression of a mutant Shc in which all three tyrosines (Tyr^{239}, Tyr^{240}, and Tyr^{317}) are mutated, acts in a dominant negative fashion to inhibit cellular mitogenesis. It is likely that angiotensin II-mediated c-Src activation may cause phosphorylation of multiple tyrosine residues on Shc, resulting in activation of divergent signaling pathways that lead to mitogenesis.

The duration and magnitude of MAP kinase activation can determine the physiological outcome within a cell. Thus, dephopshorylation of the MAP kinases is vital for their control. A plethora of studies suggest that MAP kinases are inactivated by members of the dual specificity MAP kinase phosphatase family termed, MAP kinase phosphatase (MKP). Approximately 10 members of this family have been identified (Theodosiou and Ashworth, 2002). These phosphatases have similar structural domains composed of an N-terminal non-catalytic domain containing two short sequence motifs (CH2A and CH2B) and a c-terminal catalytic domain (Keyse, 2000). The laboratory of Dr. Brad Berk was the first to report the angiotensin II-induced expression of MKP-1 in VSMC (Duff et al., 1995). Subsequently it was demonstrated that dephosphorylation of ERK by MKP-1 regulates ERK activity in VSMC (Metzler et al., 1999). Lai and co-workers (Lai et al., 1996) reported that following balloon injury, MKP-1 activity decreased, while ERK activity increased resulting in VSMC proliferation. Consistent with this observation, it was reported that mechanical stress induced activation of MKP-1 and this is associated with reduced VSMC proliferation and inactivation of ERK (Li and Xu, 2000). There are several examples of how MAP kinase phosphatases are key mediators of cross talk between distinct signaling pathways activated by angiotensin II and thus may provide mechanisms for attenuation of these signaling pathways. For example, Sandberg and colleagues determined the role of MKP-1 in agiotensin II-mediated activation of the Jak-STAT pathway (Sandberg et al., 2004). Jak2 inhibits the angiotensin II-induced activation of ERK2 via induction of MKP-1 gene expression. Angiotensin II has also been reported to induce the activation of other MKP family members. It includes the activation of MKP-2 and MKP-3 in cardiac myocytes and endothelial cells (Palm-Leis et al., 2004; Rossig et al., 2002). In cardiac myocytes, MKP-2 activation is associated with the *all trans* retinoic acid mediated anti-hypertrophy. On the other hand, MKP-3 mediates endothelial cell apoptosis. From these findings, it appears that MAP kinase phosphatases are involved in diverse functions in the context of angiotensin II signaling.

In recent years, ubiquitination and the proteasome-dependent signaling pathway have been described as another mode of regulation of the MAP kinase signaling pathway. For example, work by Lu reported that MEKK1 mediated ubiquitination and degradation of ERK1/2 (Lu et al., 2002). In contrast, it was reported that activation of ERK1/2 is associated with the degradation of MKP-1 via the ubiqutin-proteasome pathway, thus facilitating sustained activation of ERK1/2 (Lin et al., 2003). Although such a mechanism has yet to be established for AT_1 receptor signaling, the ubiquitin-proteasome pathway might be another mechanism by which the cell silences angiotensin II-mediated MAP kinase signaling. To support this hypothesis, we recently found that treatment of VSMC with MG132, a proteasome inhibitor, potentiated the extent and duration of activation of ERK1/2 in VSMC (unpublished results). This finding will be extremely useful in identifying important proteins in the regulation of angiotensin II-induced, activation of the MAP kinase pathway. Thus, further analyses addressing the regulation of these factors will be critical for the determination of the physiological regulatory processes of activation of the MAP kinase pathway in VSMC.

IV. Regulation of the Jak-STAT Signaling Pathway

The Janus kinase (Jak) family of non-receptor tyrosine kinases includes Jak1, Jak2, Jak3, Tyk2, and the Drosophila homologue hopscotch (Briscoe et al., 1996). Each protein is approximately 130 kDaltons in mass and contains seven conserved Jak homology domains. Initial review of the amino acid sequence indicated that this family of kinases contained two kinase domains and thus was named after Janus, the Roman god of two faces. Subsequent analysis revealed that the N-terminal kinase domain was inactive and was therefore termed the pseudokinase domain. Unlike almost all other protein tyrosine kinases, members of the Jak family bear no SH2 or SH3 domains. Nonetheless, Jaks are key mediators of mRNA expression. Jaks accomplish this by phosphorylating the Signal Transducers and Activators of Transcription (STATs). Phosphorylated STATs translocate to the nucleus where they bind to Sis-inducible promoter elements and stimulate transcription of the early growth response genes such as c-*fos*, c-*jun*, and c-*myc*. The increased tyrosine phosphorylation, nuclear translocation, and DNA binding activity of the STATs occurs in the presence of cycloheximide, suggesting that this signaling pathway uses a post-translational modification of existing proteins and does not require *de-novo* protein synthesis. Thus, Jaks are capable of transducing a signal from the cell surface to the nucleus through a tyrosine phosphorylation signal transduction cascade.

Studies have now shown that angiotensin II directly activates the Jak/STAT pathway via the AT_1 receptor. Initially, it was demonstrated that angiotensin II rapidly and transiently activates Jak2 and Tyk2 in VSMC (Marrero et al., 1995). In addition, STAT1, STAT2, and STAT3 are tyrosine phosphorylated in response to angiotensin II and both STAT1 and

STAT3 translocate to the nucleus. Angiotensin II also causes a transient association of Jak2 with the AT_1 receptor. Subsequent work demonstrated that this association requires amino acids 319-322 of the AT_1 receptor encoding the YIPP protein motif (Ali et al., 1997a). Similar results have been obtained in cell systems other than VSMC. A CHO cell line that stably expressed the AT_1 receptor was found to 1) activate Stat1 2) cause nuclear translocation of Stat1 and 3) increase DNA binding activity of the Stat1 in response to angiotensin II (Ali et al., 1997a; Bhat et al., 1994; Bhat et al., 1995). Furthermore, the angiotensin II-dependent activation of STAT1 and STAT3 was found to require Jak2 in neonatal rat cardiac myocytes (McWhinney et al., 1997).

In addition to angiotensin II, other GPCR ligands have been reported to activate the Jak/STAT pathway. Thrombin has been shown to increase Jak2 kinase activity and cause STAT phosphorylation (Bhat et al., 1997; Rodriguez-Linares and Watson, 1994). Activation of the ET_A receptor by endothelin also leads to STAT activation (Peeler et al., 1996). Collectively, these studies suggest that GPCR can activate the Jak/STAT pathway and thus transduce a signal from the cell surface to transcriptional events within the nucleus. The observation that Jak2 can physically associate with the AT_1 receptor also indicates that the activation of the Jak/STAT pathway is one of the earliest events in angiotensin II tyrosine phosphorylation signal transduction cascades. Targeted disruption of the *Jak2* gene results in an embryonic lethal phenotype in which myeloid and erythroid development are absent (Neubauer et al., 1998; Parganas et al., 1998). The functional consequence of Jak2 activation in VSMC is less clear and is under current investigation.

Angiotensin II binding to the AT_1 receptor leads to dissociation of the G-protein into its α and $\beta\gamma$ subunits, both of which can activate or inhibit downstream effector molecules. To study the frequently reiterated paradigm of the G-protein coupling ability of the AT_1 receptor, we examined the role of $G\alpha q$ in activation Jak-STAT pathway. Our work established that activation of the Jak-STAT pathway is independent of heterotrimeric G-proteins (Doan et al., 2001). A very exciting recent development that corroborated to our studies is that Rho small G-proteins (Rac1 and Rho A) play a novel role in the AT_1 receptor-mediated activation of the Jak-STAT pathway (Pelletier et al., 2003). Contrary to these observations, it was reported that angiotensin II-induced activation of the Jak-STAT pathway is subsequent to classical G-protein coupling. (Frank et al., 2002). The possible explanation for such a discrepancy might be cell specific effect. In VSMC the activation of STATs is transient. This suggests that efficient mechanisms for STAT inactivation must exist. Theoretically, three mechanisms can be envisioned, namely dephosphorylation, proteasome mediated degradation and inhibition by SOCS family of proteins. There is a report describing the angiotensin II mediated activation of SOCS (Hernandez-Vargas et al., 2005). These authors demonstrated that over expression of SOCS proteins prevented the STAT activation in response to angiotensin II. Antisense oligonucleotides against SOCS protein increased the angiotensin II-induced STAT activation and *c-fos/c-jun* expression. We can find no reference that describes the role of tyrosine phosphatase and proteasome mediated degradation in modulating the activity of Jak-STAT pathway in response to angiotensin II. Finally, the internalization of the AT_1 receptor is expected to be an additional tier of control on activation of the Jak-STAT pathway.

V. Regulation of Angiotensin II-Induced Signaling by Protein Tyrosine Phosphatases

A substantial amount of AT_1 receptor signaling occurs via the activation of several nonreceptor tyrosine kinases. These include Jak2, Tyk2, Src, Fyn, Pyk2, and FAK (as reviewed in Sayeski *et al.*, 1998). By phosphorylating multiple downstream substrates, these protein tyrosine kinases trigger signaling along several different pathways, leading to the activation of transcription factors and ultimately facilitate increased expression of early growth response genes. Thus, tyrosine phosphorylation represents a critical control mechanism for angiotensin II-induced mitogenesis in different cell types. Protein tyrosine phosphorylation is regulated by the opposing activities of protein tyrosine kinases and protein tyrosine phosphatases. In the past, most investigative efforts have centered on elucidating the roles of tyrosine kinases, but little attention has been devoted towards an understanding of the negative regulators that terminate angiotensin II signaling. Among these negative regulatory factors, protein tyrosine phosphatases might play an important role in preventing a potentially deleterious chronic activation of downstream signaling pathways. Among the tyrosine phosphatases activated by angiotensin II are Src homology phosphatase-1 (SHP-1), Src homology phosphatase-2 (SHP-2), PTP1B, and MAP kinase phosphatase (Ali et al., 1997b; Duff et al., 1995; Guillemot et al., 2000; Marrero et al., 2004; Shah et al., 2004). From several independent laboratories, it was reported that angiotensin II activates SHP-1 solely through the AT_2 receptor, whereas angiotensin II-dependent activation of SHP-2 is specific for the AT_1 receptor (Feng *et al.*, 2002; Shibasaki *et al.*, 2001).

SHP-2 is a ubiquitously expressed protein and critical component of several signaling pathways. It has two SH2 domains, a protein tyrosine phosphatase domain containing the signature "HCXXGXXRS(T)" sequence and a carboxyl terminus with two potential tyrosine phosphorylation sites. Crystal structure analysis revealed that the amino terminal SH2 domain is inserted into the catalytic cleft of the phosphatase domain, maintaining the enzyme in an autoinhibitory state (Barford and Neel, 1998). However, occupation of the SH2 domain by ligand binding stimulated the phosphatase activity (Feng, 1999). The net effect can be either positive or negative depending on the specific type of signaling network involved. A great deal of evidence has demonstrated that SHP-2 plays critical regulatory roles in numerous cellular events. For example, the disruption of the mouse SHP-2 gene has been shown to cause embryonic lethality (Saxton et al., 1997). Furthermore, SHP-2 has been reported to be essential for mammalian limb development (Saxton et al., 2000). Several recent reports have indicated a clinical relevance of SHP-2 to some human diseases such as Noonan syndrome (Tartaglia et al., 2001), neutropenia (Tidow et al., 1999), diabetes (Arrandale et al., 1996), and gastric ulcer and carcinoma (Higashi et al., 2002). The functional importance of SHP-2 activation in the pathophysiology of vascular diseases is still emerging. For example, Ali *et al.* (Ali et al., 1997b) demonstrated that angiotensin II induces SHP-2 tyrosine phosphorylation, activation of its phosphatase activity, and the formation of an SHP-2-containing complex with IRS-1. Subsequently, our work established that SHP-2 regulates angiotensin II-induced activation of the MAP kinase pathway by inactivating c-Src in VSMC (Doan et al., 2004). In agreement with this work, we recently established a link between the activation of SHP-2 and angiotensin II-induced actin cytoskeleton reorganization in VSMC

(Fernstrom et al., 2005). Tang and colleagues reported that angiotensin II-induced tyrosine phosphorylation of Pyk2 is regulated by SHP-2 in endothelial cells. SHP-2-mediated dephosphorylation of Pyk2 correlates with the negative effect of SHP-2 on the angiotensin II-induced ERK and JNK activation in these cells (Tang et al., 2000). Work by Guillemot demonstrated the possible involvement of SHP-2 downstream of the AT_1 receptor to control cell cycle progression (Guillemot et al., 2000). These results suggest that angiotensin II-induced activation of SHP-2 inferred a heterogenous array of signaling events.

There are two working models by which SHP-2 may regulate MAP kinase activity. One mechanism involves an SHP-2 C-terminal phosphotyrosine residue that serves as a docking site for Grb2, which then facilitates complex formation with mSos (Li et al., 1994). This potentially important post-translational modification of SHP-2 may thus allow Ras and ultimately the MAP kinase pathway to become activated. The second mechanism relies on SHP-2 phosphatase activity, and the net effect produced by this catalytic activity may be either positive or negative, depending on the signaling pathway that is affected. We have found that angiotensin II failed to induce any detectable complex formation between SHP-2 and Grb2. This observation appears to exclude the possibility that SHP-2 functions as an adaptor protein in upstream signaling that leads to activation of MAP kinase. Therefore, the biological effects of SHP-2 in angiotensin II-stimulated cells instead depends on its phosphatase activity. To elucidate the role of SHP-2 in angiotensin II-stimulated signaling, we compared tyrosine kinase activities in stably transfected RASM cells that expressed wild type or catalytically inactive SHP-2. Our results showed that SHP-2 is an efficient negative regulator of angiotensin II signaling (Doan et al., 2004). This function is associated with an ability of SHP-2 to down-regulate c-Src catalytic activity. This is supported by the observation that wild type SHP-2 promotes a significant reduction in the phosphorylation of the positive regulatory Tyr^{418} within c-Src. The consequence of the dephosphorylation of this critical tyrosine residue within the c-Src catalytic domain appears to be the inhibition of angiotensin II-induced ERK activation in VSMC. Interestingly, SHP-2 does not directly affect the angiotensin II-induced activation of the Jak-STAT pathway. These data support the concept that SHP-2 exhibits restricted substrate specificity. Considering this property, it is intriguing that SHP-2 is potentially capable of regulating complex biological phenomena that arise in response to angiotensin II stimulation.

Vascular remodeling, or the alteration of vessel structure, is one of the pathophysiological conditions promoted by angiotensin II. Rearrangements of the VSMC cytoskeleton and focal adhesions are potential key events that might contribute to vascular remodeling. The actin network is regulated by a variety of actin binding proteins, tyrosine kinases, and tyrosine phosphatases. The Src family of tyrosine kinases, focal adhesion kinase (FAK), and the Abl kinase are major tyrosine kinases that regulate cytoskeletal organization. As counterparts of tyrosine kinases, several tyrosine phosphatases have been implicated in cytoskeletal re-organization. These include PTP1B (Liu et al., 1998), SHP-1 (Roach et al., 1998), PTP-PEST (Shen et al., 1998), PTEN (Tamura et al., 1998), and SHP-2 (Yu et al., 1998). SHP-2 is an intracellular enzyme that appears unique among this family of phosphatases in that it can participate in linking proximal receptor mediated biochemical reactions to the cell movement machinery. For example, fibroblast cells isolated from $SHP\text{-}2^{-/-}$ embryos are defective in cell spreading, migration, and focal adhesion turnover (Yu et al., 1998). Furthermore, the over-expression of a catalytically inactive form of SHP-2 in CHO cells induced marked changes in cell morphology accompanied by an increase in the number of actin stress fibers and focal adhesion contacts (Inagaki et al., 2000). These results suggest that SHP-2 regulates the actin cytoskeleton organization in many types of cells. Angiotensin II has previously been reported to induce actin reorganization in a variety of cell types including VSMC, cardiac fibroblasts, and cardiac myocytes (Aikawa et al., 2000; Aoki et al., 1998; Turner et al., 1995). In contrast to protein tyrosine phosphorylation, the role of tyrosine phosphatases in angiotensin II-induced actin dynamics is poorly characterized. Thus, we hypothesize that SHP-2 regulates angiotensin II-induced actin dynamics in VSMC that have profound effects on vascular remodeling. To test this hypothesis, actin dynamics was assessed by monitoring the pattern of actin-based immunofluorescence in stably selected RASM cells which overexpress either wild type SHP-2 enzyme or a catalytically inactive SHP-2 molecule (Fernstrom et al., 2005). In the absence of angiotensin II stimulation, VSMC expressing a catalytically inactive SHP-2 (C-S) protein exhibited an increased abundance of stress fibers compared to wild type SHP-2 cells. Angiotensin II further increased the stress fiber formation within SHP-2 (C-S) expressing cells, but no new actin bundles were observed in VSMC expressing the wild type SHP-2 protein. Our results suggest that SHP-2 may negatively regulate the activity of RhoA through suppression of the GDP/GTP exchange activity of RhoA-specific Vav2. Thus, SHP-2 tyrosine phosphatase activity appears critical in linking the AT_1 receptor with a capacity to remodel the actin cytoskeleton by virtue of its ability to modulate communication along a RhoA/Vav2 signaling axis. Our data thus provide a novel mechanism by which SHP-2 regulates angiotensin II-induced actin cytoskeletal organization in VSMC.

SHP-2 is tyrosine phosphorylated at Tyr^{542} and Tyr^{580} in response to growth factor receptor activation with concomitant increases in its catalytic activity (Feng, 1999). Angiotensin II treatment of VSMC causes rapid and robust tyrosine phosphorylation of SHP-2 (Ali et al., 1997b). This would imply that tyrosine residue(s) might be responsible for angiotensin II-dependent SHP-2 catalytic activity. However, there is little evidence to suggest that SHP-2 phosphatase activity is regulated by tyrosine phosphorylation in response to angiotensin II. The classical signaling pathway for the AT_1 receptor is the heterotrimeric G-protein. Although work from several laboratories supports a critical role of SHP-2 in a variety of biological systems, its mechanism of activation, particularly the role of heterotrimeric G-proteins, has yet to be determined. Treatment of VSMC with genistein, a broad-spectrum tyrosine kinase inhibitor, completely blocked the tyrosine phosphorylation of SHP-2. This result suggests the possible involvement of tyrosine kinases in the phosphorylation of SHP-2. Thus, an unresolved issue is how upstream tyrosine kinases are responsible for the phosphorylation of SHP-2, in response to angiotensin II. A common feature in signaling pathways is the assembly of a multiprotein signaling complex that involve extensive protein-protein interactions mediated by different protein domains. The formation or assembly of multiprotein signaling complexes is necessary to transmit the signal and to elicit the appropriate subcellular compartmentalization. The molecular architecture of SHP-2 suggests that signaling proteins might interact with SHP-2 to accomplish signaling tasks within the cell. In addition, multiple signaling complexes exist at various subcellular locations that portray the multiplicity of SHP-2 function. Therefore, it is of pivotal importance to identify components of SHP-2 signaling complexes that assemble in VSMC, in response to angiotensin II.

Concluding Remarks

Angiotensin II is the effector molecule of the renin-angiotensin-aldosterone system. Its high affinity binding to the AT_1 receptor results in well established hemodynamic and electrolyte responses. The intent of this review was to demonstrate that angiotensin II is also a potent growth factor. Under certain circumstances however, the growth promoting effects of angiotensin II are not well controlled and it is these maladaptive growth responses that mediate many of the vascular pathologies seen in human disease states. In this review, we discussed the multiple signaling pathways of the AT_1 receptor. Scientific studies are currently at the crossroads of determining which of these signaling mechanisms is rate-limiting in angiotensin II signaling and more importantly, is the rate-limiting pathway altered in disease states.

Acknowledegements

This work was supported by National Institutes of Health Grant 61710 to MSA. Dr. Sayeski was supported by National Institutes of Health Awards K01-DK60471 and R01-HL67277.

References

AbdAlla, S., H. Lother, A.M. Abdel-tawab, and U. Quitterer. 2001. The angiotensin II AT2 receptor is an AT1 receptor antagonist. *J Biol Chem*. 276:39721-6.

AbdAlla, S., H. Lother, and U. Quitterer. 2000. AT1-receptor heterodimers show enhanced G-protein activation and altered receptor sequestration. *Nature*. 407:94-8.

Aikawa, R., I. Komuro, R. Nagai, and Y. Yazaki. 2000. Rho plays an important role in angiotensin II-induced hypertrophic responses in cardiac myocytes. *Mol Cell Biochem*. 212:177-82.

Ali, M.S., P.P. Sayeski, and K.E. Bernstein. 2000. Jak2 acts as both a STAT1 kinase and as a molecular bridge linking STAT1 to the angiotensin II AT1 receptor. *J Biol Chem*. 275:15586-93.

Ali, M.S., P.P. Sayeski, L.B. Dirksen, D.J. Hayzer, M.B. Marrero, and K.E. Bernstein. 1997a. Dependence on the motif YIPP for the physical association of Jak2 kinase with the intracellular carboxyl tail of the angiotensin II AT1 receptor. *J Biol Chem*. 272:23382-8.

Ali, M.S., B. Schieffer, P. Delafontaine, K.E. Bernstein, B.N. Ling, and M.B. Marrero. 1997b. Angiotensin II stimulates tyrosine phosphorylation and activation of insulin receptor substrate 1 and protein-tyrosine phosphatase 1D in vascular smooth muscle cells. *J Biol Chem*. 272:12373-9.

Angers, S., A. Salahpour, and M. Bouvier. 2002. Dimerization: an emerging concept for G protein-coupled receptor ontogeny and function. *Annu Rev Pharmacol Toxicol.* 42:409-35.

Aoki, H., S. Izumo, and J. Sadoshima. 1998. Angiotensin II activates RhoA in cardiac myocytes: a critical role of RhoA in angiotensin II-induced premyofibril formation. *Circ Res.* 82:666-76.

Arai, H., and J.A. Escobedo. 1996. Angiotensin II type 1 receptor signals through Raf-1 by a protein kinase C-dependent, Ras-independent mechanism. *Mol Pharmacol.* 50:522-8.

Arora, K.K., Z. Cheng, and K.J. Catt. 1997. Mutations of the conserved DRS motif in the second intracellular loop of the gonadotropin-releasing hormone receptor affect expression, activation, and internalization. *Mol Endocrinol.* 11:1203-12.

Arrandale, J.M., A. Gore-Willse, S. Rocks, J.M. Ren, J. Zhu, A. Davis, J.N. Livingston, and D.U. Rabin. 1996. Insulin signaling in mice expressing reduced levels of Syp. *J Biol Chem.* 271:21353-8.

Barford, D., and B.G. Neel. 1998. Revealing mechanisms for SH2 domain mediated regulation of the protein tyrosine phosphatase SHP-2. *Structure.* 6:249-54.

Barnard, E.A. 1992. Receptor classes and the transmitter-gated ion channels. *Trends Biochem Sci.* 17:368-74.

Bennett, T.A., D.C. Maestas, and E.R. Prossnitz. 2000. Arrestin binding to the G protein-coupled N-formyl peptide receptor is regulated by the conserved "DRY" sequence. *J Biol Chem.* 275:24590-4.

Bhat, G.J., S.T. Abraham, H.A. Singer, and K.M. Baker. 1997. Alpha-thrombin stimulates sis-inducing factor-A DNA binding activity in rat aortic smooth muscle cells. *Hypertension.* 29:356-60.

Bhat, G.J., T.J. Thekkumkara, W.G. Thomas, K.M. Conrad, and K.M. Baker. 1994. Angiotensin II stimulates sis-inducing factor-like DNA binding activity. Evidence that the AT1A receptor activates transcription factor-Stat91 and/or a related protein. *J Biol Chem.* 269:31443-9.

Bhat, G.J., T.J. Thekkumkara, W.G. Thomas, K.M. Conrad, and K.M. Baker. 1995. Activation of the STAT pathway by angiotensin II in T3CHO/AT1A cells. Cross-talk between angiotensin II and interleukin-6 nuclear signaling. *J Biol Chem.* 270:19059-65.

Bihoreau, C., C. Monnot, E. Davies, B. Teutsch, K.E. Bernstein, P. Corvol, and E. Clauser. 1993. Mutation of Asp74 of the rat angiotensin II receptor confers changes in antagonist affinities and abolishes G-protein coupling. *Proc Natl Acad Sci U S A.* 90:5133-7.

Briscoe, J., F. Kohlhuber, and M. Muller. 1996. JAKs and STATs branch out. *Trends Cell Biol.* 6:336-40.

Bueno, O.F., L.J. De Windt, H.W. Lim, K.M. Tymitz, S.A. Witt, T.R. Kimball, and J.D. Molkentin. 2001. The dual-specificity phosphatase MKP-1 limits the cardiac hypertrophic response in vitro and in vivo. *Circ Res.* 88:88-96.

Bueno, O.F., L.J. De Windt, K.M. Tymitz, S.A. Witt, T.R. Kimball, R. Klevitsky, T.E. Hewett, S.P. Jones, D.J. Lefer, C.F. Peng, R.N. Kitsis, and J.D. Molkentin. 2000. The MEK1-ERK1/2 signaling pathway promotes compensated cardiac hypertrophy in transgenic mice. *Embo J.* 19:6341-50.

Buggy, J.J. 1998. Binding of alpha-melanocyte-stimulating hormone to its G-protein-coupled receptor on B-lymphocytes activates the Jak/STAT pathway. *Biochem J.* 331 (Pt 1):211-6.

Burgering, B.M., and P.J. Coffer. 1995. Protein kinase B (c-Akt) in phosphatidylinositol-3-OH kinase signal transduction. *Nature.* 376:599-602.

Claing, A., S.A. Laporte, M.G. Caron, and R.J. Lefkowitz. 2002. Endocytosis of G protein-coupled receptors: roles of G protein-coupled receptor kinases and beta-arrestin proteins. *Prog Neurobiol.* 66:61-79.

Crawford, K.W., E.A. Frey, and T.E. Cote. 1992. Angiotensin II receptor recognized by DuP753 regulates two distinct guanine nucleotide-binding protein signaling pathways. *Mol Pharmacol.* 41:154-62.

Daviet, L., J.Y. Lehtonen, K. Tamura, D.P. Griese, M. Horiuchi, and V.J. Dzau. 1999. Cloning and characterization of ATRAP, a novel protein that interacts with the angiotensin II type 1 receptor. *J Biol Chem.* 274:17058-62.

de Gasparo, M., K.J. Catt, T. Inagami, J.W. Wright, and T. Unger. 2000. International union of pharmacology. XXIII. The angiotensin II receptors. *Pharmacol Rev.* 52:415-72.

Dhar, A., and S.D. Shukla. 1994. Electrotransjection of pp60v-src monoclonal antibody inhibits activation of phospholipase C in platelets. A new mechanism for platelet-activating factor responses. *J Biol Chem.* 269:9123-7.

Doan, T., P. Farmer, T. Cooney, and M.S. Ali. 2004. Selective down-regulation of angiotensin II receptor type 1A signaling by protein tyrosine phosphatase SHP-2 in vascular smooth muscle cells. *Cell Signal.* 16:301-11.

Doan, T.N., M.S. Ali, and K.E. Bernstein. 2001. Tyrosine kinase activation by the angiotensin II receptor in the absence of calcium signaling. *J Biol Chem.* 276:20954-8.

Duff, J.L., B.P. Monia, and B.C. Berk. 1995. Mitogen-activated protein (MAP) kinase is regulated by the MAP kinase phosphatase (MKP-1) in vascular smooth muscle cells. Effect of actinomycin D and antisense oligonucleotides. *J Biol Chem.* 270:7161-6.

Eguchi, S., T. Matsumoto, E.D. Motley, H. Utsunomiya, and T. Inagami. 1996. Identification of an essential signaling cascade for mitogen-activated protein kinase activation by angiotensin II in cultured rat vascular smooth muscle cells. Possible requirement of Gq-mediated p21ras activation coupled to a Ca2+/calmodulin-sensitive tyrosine kinase. *J Biol Chem.* 271:14169-75.

Erpel, T., and S.A. Courtneidge. 1995. Src family protein tyrosine kinases and cellular signal transduction pathways. *Curr Opin Cell Biol.* 7:176-82.

Feng, G.S. 1999. Shp-2 tyrosine phosphatase: signaling one cell or many. *Exp Cell Res.* 253:47-54.

Fernstrom, K. P. Farmer, and M. S. Ali. 2005. Cytoskeletal remodeling in vascular smooth muscle Cells in response to angiotensin II-induced activation of the SHP-2 tyrosine phosphatase. In Press

Frank, G.D., S. Saito, E.D. Motley, T. Sasaki, M. Ohba, T. Kuroki, T. Inagami, and S. Eguchi. 2002. Requirement of Ca(2+) and PKCdelta for Janus kinase 2 activation by angiotensin II: involvement of PYK2. *Mol Endocrinol.* 16:367-77.

Gille, H., T. Strahl, and P.E. Shaw. 1995. Activation of ternary complex factor Elk-1 by stress-activated protein kinases. *Curr Biol.* 5:1191-200.

Groblewski, T., B. Maigret, R. Larguier, C. Lombard, J.C. Bonnafous, and J. Marie. 1997. Mutation of Asn111 in the third transmembrane domain of the AT1A angiotensin II receptor induces its constitutive activation. *J Biol Chem*. 272:1822-6.

Guillemot, L., A. Levy, Z.J. Zhao, G. Bereziat, and B. Rothhut. 2000. The protein-tyrosine phosphatase SHP-2 is required during angiotensin II-mediated activation of cyclin D1 promoter in CHO-AT1A cells. *J Biol Chem*. 275:26349-58.

Guo, D.F., I. Chenier, V. Tardif, S.N. Orlov, and T. Inagami. 2003. Type 1 angiotensin II receptor-associated protein ARAP1 binds and recycles the receptor to the plasma membrane. *Biochem Biophys Res Commun*. 310:1254-65.

Hanke, J.H., J.P. Gardner, R.L. Dow, P.S. Changelian, W.H. Brissette, E.J. Weringer, B.A. Pollok, and P.A. Connelly. 1996. Discovery of a novel, potent, and Src family-selective tyrosine kinase inhibitor. Study of Lck- and FynT-dependent T cell activation. *J Biol Chem*. 271:695-701.

Hansen, J.L., J. Theilade, S. Haunso, and S.P. Sheikh. 2004. Oligomerization of wild type and nonfunctional mutant angiotensin II type I receptors inhibits galphaq protein signaling but not ERK activation. *J Biol Chem*. 279:24108-15.

He, T.C., N. Jiang, H. Zhuang, and D.M. Wojchowski. 1995. Erythropoietin-induced recruitment of Shc via a receptor phosphotyrosine-independent, Jak2-associated pathway. *J Biol Chem*. 270:11055-61.

Hernandez-Vargas, P., O. Lopez-Franco, G. Sanjuan, M. Ruperez, G. Ortiz-Munoz, Y. Suzuki, P. Aguado-Roncero, G. Perez-Tejerizo, J. Blanco, J. Egido, M. Ruiz-Ortega, and C. Gomez-Guerrero. 2005. Suppressors of Cytokine Signaling Regulate Angiotensin II-Activated Janus Kinase-Signal Transducers and Activators of Transcription Pathway in Renal Cells. *J Am Soc Nephrol*.

Hernanz-Falcon, P., J.M. Rodriguez-Frade, A. Serrano, D. Juan, A. del Sol, S.F. Soriano, F. Roncal, L. Gomez, A. Valencia, A.C. Martinez, and M. Mellado. 2004. Identification of amino acid residues crucial for chemokine receptor dimerization. *Nat Immunol*. 5:216-23.

Higashi, H., R. Tsutsumi, S. Muto, T. Sugiyama, T. Azuma, M. Asaka, and M. Hatakeyama. 2002. SHP-2 tyrosine phosphatase as an intracellular target of Helicobacter pylori CagA protein. *Science*. 295:683-6.

Hines, J., S.J. Fluharty, and D.K. Yee. 2003. Structural determinants for the activation mechanism of the angiotensin II type 1 receptor differ for phosphoinositide hydrolysis and mitogen-activated protein kinase pathways. *Biochem Pharmacol*. 66:251-62.

Hunyady, L., M. Bor, T. Balla, and K.J. Catt. 1994. Identification of a cytoplasmic Ser-Thr-Leu motif that determines agonist-induced internalization of the AT1 angiotensin receptor. *J Biol Chem*. 269:31378-82.

Inada, Y., T. Nakane, and S. Chiba. 2003. Relationship between ligand binding and YIPP motif in the C-terminal region of human AT1 receptor. *Biochim Biophys Acta*. 1640:33-41.

Inagaki, K., T. Noguchi, T. Matozaki, T. Horikawa, K. Fukunaga, M. Tsuda, M. Ichihashi, and M. Kasuga. 2000. Roles for the protein tyrosine phosphatase SHP-2 in cytoskeletal organization, cell adhesion and cell migration revealed by overexpression of a dominant negative mutant. *Oncogene*. 19:75-84.

Inagami, T., N. Iwai, K. Sasaki, Y. Yamamo, S. Bardhan, S. Chaki, D.F. Guo, and H. Furuta. 1992. Cloning, expression and regulation of angiotensin II receptors. *J Hypertens.* 10:713-6.

Inoue, Y., N. Nakamura, and T. Inagami. 1997. A review of mutagenesis studies of angiotensin II type 1 receptor, the three-dimensional receptor model in search of the agonist and antagonist binding site and the hypothesis of a receptor activation mechanism. *J Hypertens.* 15:703-14.

Ju, H., V.J. Venema, M.B. Marrero, and R.C. Venema. 1998. Inhibitory interactions of the bradykinin B2 receptor with endothelial nitric-oxide synthase. *J Biol Chem.* 273:24025-9.

Kai, H., R.W. Alexander, M. Ushio-Fukai, P.R. Lyons, M. Akers, and K.K. Griendling. 1998. G-Protein binding domains of the angiotensin II AT1A receptors mapped with synthetic peptides selected from the receptor sequence. *Biochem J.* 332 (Pt 3):781-7.

Kai, H., T. Fukui, B. Lassegue, A. Shah, C.A. Minieri, and K.K. Griendling. 1996. Prolonged exposure to agonist results in a reduction in the levels of the Gq/G11 alpha subunits in cultured vascular smooth muscle cells. *Mol Pharmacol.* 49:96-104.

Kai, H., K.K. Griendling, B. Lassegue, J.D. Ollerenshaw, M.S. Runge, and R.W. Alexander. 1994. Agonist-induced phosphorylation of the vascular type 1 angiotensin II receptor. *Hypertension.* 24:523-7.

Keyse, S.M. 2000. Protein phosphatases and the regulation of mitogen-activated protein kinase signalling. *Curr Opin Cell Biol.* 12:186-92.

Kim, J., S. Ahn, X.R. Ren, E.J. Whalen, E. Reiter, H. Wei, and R.J. Lefkowitz. 2005. Functional antagonism of different G protein-coupled receptor kinases for beta-arrestin-mediated angiotensin II receptor signaling. *Proc Natl Acad Sci U S A.* 102:1442-7.

Lai, K., H. Wang, W.S. Lee, M.K. Jain, M.E. Lee, and E. Haber. 1996. Mitogen-activated protein kinase phosphatase-1 in rat arterial smooth muscle cell proliferation. *J Clin Invest.* 98:1560-7.

Lai, K.M., and T. Pawson. 2000. The ShcA phosphotyrosine docking protein sensitizes cardiovascular signaling in the mouse embryo. *Genes Dev.* 14:1132-45.

Leonard, W.J., and J.J. O'Shea. 1998. Jaks and STATs: biological implications. *Annu Rev Immunol.* 16:293-322.

Leszczynski, D., S. Joenvaara, and M.L. Foegh. 1996. Protein kinase C-alpha regulates proliferation but not apoptosis in rat coronary vascular smooth muscle cells. *Life Sci.* 58:599-606.

Li, C., and Q. Xu. 2000. Mechanical stress-initiated signal transductions in vascular smooth muscle cells. *Cell Signal.* 12:435-45.

Li, W., R. Nishimura, A. Kashishian, A.G. Batzer, W.J. Kim, J.A. Cooper, and J. Schlessinger. 1994. A new function for a phosphotyrosine phosphatase: linking GRB2-Sos to a receptor tyrosine kinase. *Mol Cell Biol.* 14:509-17.

Lin, Y.W., S.M. Chuang, and J.L. Yang. 2003. ERK1/2 achieves sustained activation by stimulating MAPK phosphatase-1 degradation via the ubiquitin-proteasome pathway. *J Biol Chem.* 278:21534-41.

Linseman, D.A., C.W. Benjamin, and D.A. Jones. 1995. Convergence of angiotensin II and platelet-derived growth factor receptor signaling cascades in vascular smooth muscle cells. *J Biol Chem*. 270:12563-8.

Liu, F., M.A. Sells, and J. Chernoff. 1998. Protein tyrosine phosphatase 1B negatively regulates integrin signaling. *Curr Biol*. 8:173-6.

Lopez-Ilasaca, M., X. Liu, K. Tamura, and V.J. Dzau. 2003. The angiotensin II type I receptor-associated protein, ATRAP, is a transmembrane protein and a modulator of angiotensin II signaling. *Mol Biol Cell*. 14:5038-50.

Lu, Z., S. Xu, C. Joazeiro, M.H. Cobb, and T. Hunter. 2002. The PHD domain of MEKK1 acts as an E3 ubiquitin ligase and mediates ubiquitination and degradation of ERK1/2. *Mol Cell*. 9:945-56.

Luttrell, L.M., S.S. Ferguson, Y. Daaka, W.E. Miller, S. Maudsley, G.J. Della Rocca, F. Lin, H. Kawakatsu, K. Owada, D.K. Luttrell, M.G. Caron, and R.J. Lefkowitz. 1999. Beta-arrestin-dependent formation of beta2 adrenergic receptor-Src protein kinase complexes. *Science*. 283:655-61.

Macrez-Lepretre, N., F. Kalkbrenner, J.L. Morel, G. Schultz, and J. Mironneau. 1997. G protein heterotrimer Galpha13beta1gamma3 couples the angiotensin AT1A receptor to increases in cytoplasmic Ca2+ in rat portal vein myocytes. *J Biol Chem*. 272:10095-102.

Marie, J., B. Maigret, M.P. Joseph, R. Larguier, S. Nouet, C. Lombard, and J.C. Bonnafous. 1994. Tyr292 in the seventh transmembrane domain of the AT1A angiotensin II receptor is essential for its coupling to phospholipase C. *J Biol Chem*. 269:20815-8.

Marrero, M.B., D. Fulton, D. Stepp, and D.M. Stern. 2004. Angiotensin II-induced insulin resistance and protein tyrosine phosphatases. *Arterioscler Thromb Vasc Biol*. 24:2009-13.

Marrero, M.B., B. Schieffer, W.G. Paxton, L. Heerdt, B.C. Berk, P. Delafontaine, and K.E. Bernstein. 1995. Direct stimulation of Jak/STAT pathway by the angiotensin II AT1 receptor. *Nature*. 375:247-50.

Marrero, M.B., V.J. Venema, H. Ju, D.C. Eaton, and R.C. Venema. 1998. Regulation of angiotensin II-induced JAK2 tyrosine phosphorylation: roles of SHP-1 and SHP-2. *Am J Physiol*. 275:C1216-23.

McGlade, J., A. Cheng, G. Pelicci, P.G. Pelicci, and T. Pawson. 1992. Shc proteins are phosphorylated and regulated by the v-Src and v-Fps protein-tyrosine kinases. *Proc Natl Acad Sci U S A*. 89:8869-73.

McWhinney, C.D., R.A. Hunt, K.M. Conrad, D.E. Dostal, and K.M. Baker. 1997. The type I angiotensin II receptor couples to Stat1 and Stat3 activation through Jak2 kinase in neonatal rat cardiac myocytes. *J Mol Cell Cardiol*. 29:2513-24.

Metzler, B., C. Li, Y. Hu, G. Sturm, N. Ghaffari-Tabrizi, and Q. Xu. 1999. LDL stimulates mitogen-activated protein kinase phosphatase-1 expression, independent of LDL receptors, in vascular smooth muscle cells. *Arterioscler Thromb Vasc Biol*. 19:1862-71.

Meydan, N., T. Grunberger, H. Dadi, M. Shahar, E. Arpaia, Z. Lapidot, J.S. Leeder, M. Freedman, A. Cohen, A. Gazit, A. Levitzki, and C.M. Roifman. 1996. Inhibition of acute lymphoblastic leukaemia by a Jak-2 inhibitor. *Nature*. 379:645-8.

Migliaccio, E., M. Giorgio, S. Mele, G. Pelicci, P. Reboldi, P.P. Pandolfi, L. Lanfrancone, and P.G. Pelicci. 1999. The p66shc adaptor protein controls oxidative stress response and life span in mammals. *Nature*. 402:309-13.

Miura, S., J. Zhang, J. Boros, and S.S. Karnik. 2003. TM2-TM7 interaction in coupling movement of transmembrane helices to activation of the angiotensin II type-1 receptor. *J Biol Chem*. 278:3720-5.

Murphy, T.J., R.W. Alexander, K.K. Griendling, M.S. Runge, and K.E. Bernstein. 1991. Isolation of a cDNA encoding the vascular type-1 angiotensin II receptor. *Nature*. 351:233-6.

Neubauer, H., A. Cumano, M. Muller, H. Wu, U. Huffstadt, and K. Pfeffer. 1998. Jak2 deficiency defines an essential developmental checkpoint in definitive hematopoiesis. *Cell*. 93:397-409.

Ohyama, K., Y. Yamano, T. Sano, Y. Nakagomi, M. Wada, and T. Inagami. 2002. Role of the conserved DRY motif on G protein activation of rat angiotensin II receptor type 1A. *Biochem Biophys Res Commun*. 292:362-7.

Olivares-Reyes, J.A., R.D. Smith, L. Hunyady, B.H. Shah, and K.J. Catt. 2001. Agonist-induced signaling, desensitization, and internalization of a phosphorylation-deficient AT1A angiotensin receptor. *J Biol Chem*. 276:37761-8.

Palm-Leis, A., U.S. Singh, B.S. Herbelin, G.D. Olsovsky, K.M. Baker, and J. Pan. 2004. Mitogen-activated protein kinases and mitogen-activated protein kinase phosphatases mediate the inhibitory effects of all-trans retinoic acid on the hypertrophic growth of cardiomyocytes. *J Biol Chem*. 279:54905-17.

Parganas, E., D. Wang, D. Stravopodis, D.J. Topham, J.C. Marine, S. Teglund, E.F. Vanin, S. Bodner, O.R. Colamonici, J.M. van Deursen, G. Grosveld, and J.N. Ihle. 1998. Jak2 is essential for signaling through a variety of cytokine receptors. *Cell*. 93:385-95.

Pawson, T. 1997. New impressions of Src and Hck. *Nature*. 385:582-3, 585.

Paxton, W.G., M.B. Marrero, J.D. Klein, P. Delafontaine, B.C. Berk, and K.E. Bernstein. 1994. The angiotensin II AT1 receptor is tyrosine and serine phosphorylated and can serve as a substrate for the src family of tyrosine kinases. *Biochem Biophys Res Commun*. 200:260-7.

Peeler, T.C., K.M. Conrad, and K.M. Baker. 1996. Endothelin stimulates sis-inducing factor-like DNA binding activity in CHO-K1 cells expressing ETA receptors. *Biochem Biophys Res Commun*. 221:62-6.

Pelletier, S., F. Duhamel, P. Coulombe, M.R. Popoff, and S. Meloche. 2003. Rho family GTPases are required for activation of Jak/STAT signaling by G protein-coupled receptors. *Mol Cell Biol*. 23:1316-33.

Roach, T.I., S.E. Slater, L.S. White, X. Zhang, P.W. Majerus, E.J. Brown, and M.L. Thomas. 1998. The protein tyrosine phosphatase SHP-1 regulates integrin-mediated adhesion of macrophages. *Curr Biol*. 8:1035-8.

Rodriguez-Linares, B., and S.P. Watson. 1994. Phosphorylation of JAK2 in thrombin-stimulated human platelets. *FEBS Lett*. 352:335-8.

Rossig, L., C. Hermann, J. Haendeler, B. Assmus, A.M. Zeiher, and S. Dimmeler. 2002. Angiotensin II-induced upregulation of MAP kinase phosphatase-3 mRNA levels mediates endothelial cell apoptosis. *Basic Res Cardiol*. 97:1-8.

Sadoshima, J., and S. Izumo. 1996. The heterotrimeric G q protein-coupled angiotensin II receptor activates p21 ras via the tyrosine kinase-Shc-Grb2-Sos pathway in cardiac myocytes. *Embo J.* 15:775-87.

Sandberg, E.M., X. Ma, D. VonDerLinden, M.D. Godeny, and P.P. Sayeski. 2004. Jak2 tyrosine kinase mediates angiotensin II-dependent inactivation of ERK2 via induction of mitogen-activated protein kinase phosphatase 1. *J Biol Chem.* 279:1956-67.

Sano, T., K. Ohyama, Y. Yamano, Y. Nakagomi, S. Nakazawa, M. Kikyo, H. Shirai, J.S. Blank, J.H. Exton, and T. Inagami. 1997. A domain for G protein coupling in carboxyl-terminal tail of rat angiotensin II receptor type 1A. *J Biol Chem.* 272:23631-6.

Sasaguri, T., H. Teruya, A. Ishida, T. Abumiya, and J. Ogata. 2000. Linkage between alpha(1) adrenergic receptor and the Jak/STAT signaling pathway in vascular smooth muscle cells. *Biochem Biophys Res Commun.* 268:25-30.

Sasaki, K., Y. Yamano, S. Bardhan, N. Iwai, J.J. Murray, M. Hasegawa, Y. Matsuda, and T. Inagami. 1991. Cloning and expression of a complementary DNA encoding a bovine adrenal angiotensin II type-1 receptor. *Nature.* 351:230-3.

Sasaoka, T., D.W. Rose, B.H. Jhun, A.R. Saltiel, B. Draznin, and J.M. Olefsky. 1994. Evidence for a functional role of Shc proteins in mitogenic signaling induced by insulin, insulin-like growth factor-1, and epidermal growth factor. *J Biol Chem.* 269:13689-94.

Sato, K., T. Otsuki, M. Kimoto, M. Kakumoto, A.A. Tokmakov, Y. Watanabe, and Y. Fukami. 1998. c-Src and phosphatidylinositol 3-kinase are involved in NGF-dependent tyrosine phosphorylation of Shc in PC12 cells. *Biochem Biophys Res Commun.* 250:223-8.

Saxton, T.M., B.G. Ciruna, D. Holmyard, S. Kulkarni, K. Harpal, J. Rossant, and T. Pawson. 2000. The SH2 tyrosine phosphatase shp2 is required for mammalian limb development. *Nat Genet.* 24:420-3.

Saxton, T.M., M. Henkemeyer, S. Gasca, R. Shen, D.J. Rossi, F. Shalaby, G.S. Feng, and T. Pawson. 1997. Abnormal mesoderm patterning in mouse embryos mutant for the SH2 tyrosine phosphatase Shp-2. *Embo J.* 16:2352-64.

Sayeski, P.P., and M.S. Ali. 2003. The critical role of c-Src and the Shc/Grb2/ERK2 signaling pathway in angiotensin II-dependent VSMC proliferation. *Exp Cell Res.* 287:339-49.

Sayeski, P.P., M.S. Ali, D.J. Semeniuk, T.N. Doan, and K.E. Bernstein. 1998. Angiotensin II signal transduction pathways. *Regul Pept.* 78:19-29.

Schaeffer, H.J., and M.J. Weber. 1999. Mitogen-activated protein kinases: specific messages from ubiquitous messengers. *Mol Cell Biol.* 19:2435-44.

Schieffer, B., W.G. Paxton, Q. Chai, M.B. Marrero, and K.E. Bernstein. 1996. Angiotensin II controls p21ras activity via pp60c-src. *J Biol Chem.* 271:10329-33.

Schorb, W., T.C. Peeler, N.N. Madigan, K.M. Conrad, and K.M. Baker. 1994. Angiotensin II-induced protein tyrosine phosphorylation in neonatal rat cardiac fibroblasts. *J Biol Chem.* 269:19626-32.

Seta, K., M. Nanamori, J.G. Modrall, R.R. Neubig, and J. Sadoshima. 2002. AT1 receptor mutant lacking heterotrimeric G protein coupling activates the Src-Ras-ERK pathway without nuclear translocation of ERKs. *J Biol Chem.* 277:9268-77.

Seta, K., and J. Sadoshima. 2003. Phosphorylation of tyrosine 319 of the angiotensin II type 1 receptor mediates angiotensin II-induced trans-activation of the epidermal growth factor receptor. *J Biol Chem*. 278:9019-26.

Shah, B.H., A. Yesilkaya, J.A. Olivares-Reyes, H.D. Chen, L. Hunyady, and K.J. Catt. 2004. Differential pathways of angiotensin II-induced extracellularly regulated kinase 1/2 phosphorylation in specific cell types: role of heparin-binding epidermal growth factor. *Mol Endocrinol*. 18:2035-48.

Shen, Y., G. Schneider, J.F. Cloutier, A. Veillette, and M.D. Schaller. 1998. Direct association of protein-tyrosine phosphatase PTP-PEST with paxillin. *J Biol Chem*. 273:6474-81.

Shibata, T., C. Suzuki, J. Ohnishi, K. Murakami, and H. Miyazaki. 1996. Identification of regions in the human angiotensin II receptor type 1 responsible for Gi and Gq coupling by mutagenesis study. *Biochem Biophys Res Commun*. 218:383-9.

Shirai, H., K. Takahashi, T. Katada, and T. Inagami. 1995. Mapping of G protein coupling sites of the angiotensin II type 1 receptor. *Hypertension*. 25:726-30.

Tamura, M., J. Gu, K. Matsumoto, S. Aota, R. Parsons, and K.M. Yamada. 1998. Inhibition of cell migration, spreading, and focal adhesions by tumor suppressor PTEN. *Science*. 280:1614-7.

Tang, H., Z.J. Zhao, E.J. Landon, and T. Inagami. 2000. Regulation of calcium-sensitive tyrosine kinase Pyk2 by angiotensin II in endothelial cells. Roles of Yes tyrosine kinase and tyrosine phosphatase SHP-2. *J Biol Chem*. 275:8389-96.

Tartaglia, M., E.L. Mehler, R. Goldberg, G. Zampino, H.G. Brunner, H. Kremer, I. van der Burgt, A.H. Crosby, A. Ion, S. Jeffery, K. Kalidas, M.A. Patton, R.S. Kucherlapati, and B.D. Gelb. 2001. Mutations in PTPN11, encoding the protein tyrosine phosphatase SHP-2, cause Noonan syndrome. *Nat Genet*. 29:465-8.

Taubman, M.B., B.C. Berk, S. Izumo, T. Tsuda, R.W. Alexander, and B. Nadal-Ginard. 1989. Angiotensin II induces c-fos mRNA in aortic smooth muscle. Role of Ca2+ mobilization and protein kinase C activation. *J Biol Chem*. 264:526-30.

Terada, Y., K. Tomita, M.K. Homma, H. Nonoguchi, T. Yang, T. Yamada, Y. Yuasa, E.G. Krebs, and F. Marumo. 1995. Sequential activation of MAP kinase cascade by angiotensin II in opossum kidney cells. *Kidney Int*. 48:1801-9.

Theodosiou, A., and A. Ashworth. 2002. MAP kinase phosphatases. *Genome Biol*. 3:REVIEWS3009.

Thomas, W.G., H. Qian, C.S. Chang, and S. Karnik. 2000. Agonist-induced phosphorylation of the angiotensin II (AT(1A)) receptor requires generation of a conformation that is distinct from the inositol phosphate-signaling state. *J Biol Chem*. 275:2893-900.

Tian, Y., R.D. Smith, T. Balla, and K.J. Catt. 1998. Angiotensin II activates mitogen-activated protein kinase via protein kinase C and Ras/Raf-1 kinase in bovine adrenal glomerulosa cells. *Endocrinology*. 139:1801-9.

Tidow, N., B. Kasper, and K. Welte. 1999. SH2-containing protein tyrosine phosphatases SHP-1 and SHP-2 are dramatically increased at the protein level in neutrophils from patients with severe congenital neutropenia (Kostmann's syndrome). *Exp Hematol*. 27:1038-45.

Turner, C.E., K.M. Pietras, D.S. Taylor, and C.J. Molloy. 1995. Angiotensin II stimulation of rapid paxillin tyrosine phosphorylation correlates with the formation of focal adhesions in rat aortic smooth muscle cells. *J Cell Sci.* 108 (Pt 1):333-42.

van Biesen, T., L.M. Luttrell, B.E. Hawes, and R.J. Lefkowitz. 1996. Mitogenic signaling via G protein-coupled receptors. *Endocr Rev.* 17:698-714.

Venema, R.C., H. Ju, V.J. Venema, B. Schieffer, J.B. Harp, B.N. Ling, D.C. Eaton, and M.B. Marrero. 1998. Angiotensin II-induced association of phospholipase Cgamma1 with the G-protein-coupled AT1 receptor. *J Biol Chem.* 273:7703-8.

Wang, C., S. Jayadev, and J.A. Escobedo. 1995. Identification of a domain in the angiotensin II type 1 receptor determining Gq coupling by the use of receptor chimeras. *J Biol Chem.* 270:16677-82.

Wei, H., S. Ahn, W.G. Barnes, and R.J. Lefkowitz. 2004. Stable interaction between beta-arrestin 2 and angiotensin type 1A receptor is required for beta-arrestin 2-mediated activation of extracellular signal-regulated kinases 1 and 2. *J Biol Chem.* 279:48255-61.

Yin, G., J. Haendeler, C. Yan, and B.C. Berk. 2004. GIT1 functions as a scaffold for MEK1-extracellular signal-regulated kinase 1 and 2 activation by angiotensin II and epidermal growth factor. *Mol Cell Biol.* 24:875-85.

Yu, D.H., C.K. Qu, O. Henegariu, X. Lu, and G.S. Feng. 1998. Protein-tyrosine phosphatase Shp-2 regulates cell spreading, migration, and focal adhesion. *J Biol Chem.* 273:21125-31.

Zeng, F.Y., and J. Wess. 1999. Identification and molecular characterization of m3 muscarinic receptor dimers. *J Biol Chem.* 274:19487-97.

Zhang, L., V. Camerini, T.P. Bender, and K.S. Ravichandran. 2002. A nonredundant role for the adapter protein Shc in thymic T cell development. *Nat Immunol.* 3:749-55.

Zhong, H., T.J. Murphy, and K.P. Minneman. 2000. Activation of signal transducers and activators of transcription by alpha(1A)-adrenergic receptor stimulation in PC12 cells. *Mol Pharmacol.* 57:961-7.

Zhu, S.Z., S.Z. Wang, J. Hu, and E.E. el-Fakahany. 1994. An arginine residue conserved in most G protein-coupled receptors is essential for the function of the m1 muscarinic receptor. *Mol Pharmacol.* 45:517-23.

Zou, Y., I. Komuro, T. Yamazaki, R. Aikawa, S. Kudoh, I. Shiojima, Y. Hiroi, T. Mizuno, and Y. Yazaki. 1996. Protein kinase C, but not tyrosine kinases or Ras, plays a critical role in angiotensin II-induced activation of Raf-1 kinase and extracellular signal-regulated protein kinases in cardiac myocytes. *J Biol Chem.* 271:33592-7.

In: Focus on Cellular Signalling Research
Editor: Dorothy T. Leeds, pp. 29-54
ISBN 1-59454-619-3

Chapter II

Microtubules Network in Glucocorticoid Receptor and Aryl Hydrocarbon Receptor Signalling

Zdeněk Dvořák[1†], Martin Modrianský[1†], Jean-Marc Pascussi[2], Marie-Jose Vilarem[2], Patrick Maurel[2], and Jitka Ulrichova[1]
[1]Institute of Medical Chemistry and Biochemistry, Faculty of Medicine, Palacky University Olomouc, Hnevotinska 3, 77515 Olomouc, Czech Republic
[2]INSERM – UMR632, 1919 Route de Mende, 34293 Montpellier, France
[#]The authors are thankful for financial support for the grant MSM 6198959216
[†]These two authors contributed equally.

Abstract

Glucocorticoid receptor (GR) and aryl hydrocarbon receptor (AhR) are two entirely different individuals, the former being a typical steroid receptor integrated in endogenous affairs, the latter being the first described xeno-receptor, so-called dioxin receptor. Interestingly, these two receptors share several common features: 1) involvement in essential cellular processes such as proliferation, differentiation, inflammatory response, and gene expression; 2) essential role in drug metabolism, i.e. direct and/or indirect control of phase I, II, and III enzymes expression. 3) mechanism of their action in terms of ligand binding, nuclear import and export, and proteasome-mediated degradation, the last feature serving as negative feedback. 4) The dependence of their activity on the status of microtubules network. Despite voluminous data published, the role of microtubules in GR signaling is still controversial. Crucial appears the distinction between proliferating versus non-proliferating cells. While perturbation of microtubules in human cervix carcinoma cells (HeLa) resulted in rapid time- and dose-dependent, proteasome-mediated degradation of GR protein, microtubules disruption in primary human hepatocytes down-regulated GR driven genes by a mechanism involving inhibition of GR nuclear import. The mentioned similarities between GR and AhR also raise the question of the role of microtubules in AhR signaling. Nocodazole treatment caused inhibition of CYP1A1 dioxin-dependent inducibility in murine hepatoma cells

(Hepa-1), but other data regarding microtubules integrity influence on AhR activity are much more scarce. Another common denominator for AhR and GR signaling is the ambiguity whether the alteration of either receptor transcriptional activity is due to cell cycle arrest or due to direct involvement of microtubules integrity in AhR and GR signal transduction. This chapter is a summary of the evidence on the AhR and GR signaling under the microtubules network perturbation and formulates a hypothesis on the physiological role of both receptors, with regard to the implications in drug metabolism.

Key words: Glucocorticoid Receptor; Aryl Hydrocarbon Receptor; Microtubules; Cellular Signaling; Drug Metabolism

1. Introduction

Nuclear receptors are a family of distantly related regulatory proteins responsible for protein expression modulation. These proteins have certain similarities in terms of subcellular localization, translocation requirements, DNA binding site sequences and co-activators. While endogenous ligands are known for some, others remain „orphans“ [1]. A subordination appears to exist between certain members [2] with possible as yet unidentified participants with a say in the matter constituting a tangle of receptor network within a particular cell type. This network is too vast to be engaged in a single chapter, therefore we shall focus on receptors involved in xenobiotic metabolism, and even then on receptors wielding a certain superiority.

From the point of view of contemporary medicine, i.e. extensive drug use for treatment of illnesses, only a limited number of receptors are involved in xenobiotic metabolism. Furthermore, receptors actively participating in drug metabolizing genes expression in liver, and to limited extent in intestines, are subjects of interest. To date, huge knowledge is available on the role of so-called "xenoreceptors" in drug metabolism. In particular, pregnane X receptor (PXR), constitutive androstane receptor (CAR), aryl hydrocarbon receptor (AhR) were investigated [1]. Receptors CAR and PXR recruit in the process of their activation other nuclear receptors like retionic X receptor (RXR) as heterodimerization partners. The role of other nuclear receptors in the biotransformation enzymes expression still is under intense scrutiny. Namely farnesyl X receptor (FXR), lanosterol X receptor (LXR) and vitamin D receptor (VDR) were found to influence drug metabolizing enzymes expression [1]. The functional and transcriptional cross-talk between typical xenoreceptors and well characterized endogenous receptors was described [3].

Nuclear receptors comprise several families of receptors, where steroid-thyroid receptors represent a prominent group. Androstane receptor (AR), estrogen receptor (ER), glucocorticoid receptor (GR), mineralocorticoid receptor (MR), and progesterone receptor (PR) are the principal constituents of the steroid receptors family [4]. According to their physiological role, their activity is bound to the sex hormones (AR, ER, PR) or corticoids (GR, MR).

2. Methodology in the Studies on AhR and GR

This chapter intends to describe in brief the methodology commonly employed in the molecular biology and cellular signaling research with respect to AhR and GR receptors. The description is not exhaustive; however, it should serve as a cursory guide for researchers.

2.1. Gene Expression

The transmission of genetic information flowing in the order deoxyribonucleic acid (DNA) – messenger ribonucleic acid (mRNA) – protein constitutes the fundamental dogma of current molecular biology. When we simplify our thoughts, the basic investigation is focused on the determination of particular mRNA and protein content within a cell, tissue or organ. In the studies on AhR and GR receptors, we are engaged in the determination of the levels of mRNAs and proteins for: 1) AhR and GR receptors; 2) AhR and GR driven genes; 3) factors affecting AhR and GR level and activity.

2.1.1. Detection of Messenger RNA

The methods of mRNA detection/determination face the problems of sensitivity, specificity, and stability [5]. The most commonly used methods of mRNA detection are Northern blot (NB), ribonuclease protection assay (RPA) and real-time polymerase chain reaction (RT-PCR).

Northern blot - Because of lower sensitivity, it is limited to high abundance mRNA [6]. On the other hand, shift from low to high abundance (inducibility) of a gene can be easily demonstrated because of this. The receptors mRNAs are expressed at low levels, therefore this technique is recommended rather for the studies on receptor driven-genes (e.g. CYP1A1 by AhR). NB technique allows to detect two or more genes in one hybridization step and/or to perform additional detection after former RNA-DNA complex deshybridization.

Ribonuclease protection assay - This method is used for detection and, to some extent, to quantification of mRNA [7]. Owing to its unique principle, the method gives nearly absolute specificity. Compared to the NB, RPA is approximately ten times more sensitive. Its full potential lies in the possibility to map RNA termini, to determine alternative splicing and even single base mutations [8, 9].

Real time polymerase chain reaction – An elegant, safe, and simple method to use, it provides 10 000 times higher sensitivity than NB and is the only RNA detection method allowing quantification [10, 11]. Owing to its high sensitivity, a large number of genes can be analyzed from a single mRNA preparation. Other frequent uses include: primary validation of gene expression and hence confirmation of microarray data, splice variant analysis, allelic discrimination, and single nucleotide polymorphism genotyping [12].

2.1.2. Detection of Proteins

Western blot is a widespread immunodetection technique for proteins analysis. It is used not only for single protein detection and semiquantitative determination of its amount, but covalent modifications of the proteins, ability of protein to bind DNA (see bellow – EMSA), the formation of quaternary structures e.g. dimers may be evaluated. Proteolytic activation or degradation can be studied by this technique as well (vide infra). Most of the analysis is performed under denaturing conditions where only covalent bonds within the analyte are conserved. Non-covalent interactions and disulfide bonds are disrupted by boiling, reducing agents (dithiothreitol, mercaptoethanol), and anionic surfactants (sodium dodecylsulfate). Protein samples are subjected to the sodium dodecylsulfate polyacrylamide gel electrophoresis (SDS-PAGE) [13]. However, for the studies on protein-protein or protein-DNA interactions, western blots are performed under non-denaturing conditions. This technique that uses so called "native gels" has a great importance in the studies of receptors interactions with themselves (dimerization), scaffold proteins, co-activators, repressors, and DNA. The density and the size of the gel are chosen based on the molecular weight of the protein of interest, regarding the needs for particular analysis. Following the separation, the proteins are transferred onto the nitrocellulose or polyvinylidene difluoride (PVDF) membrane [14]. The proteins are then detected using two-step immune-chemiluminiscence method, where the chemiluminiscence is recorded by autoradiography.

In the past few years the technique of two-dimensional electrophoresis is of growing importance as the powerful tool in proteomic studies, including those oriented towards receptors [15].

2.2. Receptor Activation Assays

The process of AhR and GR receptors activation is very complex, consisting of ligand binding to the receptor, dissociation of scaffold proteins, nuclear import of the receptor, recruitment of co-activators and transcription factors, DNA binding, nuclear export and proteasome degradation of the receptor. Several methods are routinely used in the studies on the receptors activation

2.2.1. Ligand Binding Assay

The primary event in the process of AhR and GR activation is the binding of respective ligand to the receptor, followed by conformational change of the receptor. Ligand binding assay is the convenient and suitable method for the pharmacological studies of antagonism. As the source of AhR or GR protein for the assay are commonly used the cytosols from cells (e.g. simian kidney cells COS-1) transiently transfected with the GR or AhR expressing plasmid. The assay is based on the competition between the binding of radio-labeled ligand (i.e. [^{3}H]-dexamethasone in case of GR) and the investigated substance. Non-labeled ligand serves as the positive control [15]. Following the incubation period, bound and free ligands are separated by gel filtration and the radioactivity is measured in a liquid scintillation counter.

2.2.2. GST Pull-Down Assay

Glutathione-S-transferase (GST) pull-down assay is a variation on ligand binding assay, allowing preferably the screening on the receptors ligands/activators. The basic components needed for the assay are: 1) Radio-labeled (usually ^{35}S-Methionine) receptor; 2) Protein chimera obtained by fusion of glutathione-S-transferase and the coactivator/factor that interacts with the respective receptor. Steroid receptors co-activator 1 (SRC-1) is commonly used in the studies of nuclear hormone receptors [16, 17]. The assay can be used reciprocally, i.e. combining the radio-labeled receptor, its ligand and the fusion protein GST/X, where X stands for the protein/factor under the investigation. For instance, GST/SMRT (silencing mediator of retinoic acid and thyroid hormone receptors) protein was used to prove the interaction between AhR and SMRT [18]. The principle of the assay is ingenious. The mixture of radio-labeled receptor, appropriate fusion protein and tested substance is subjected to the column affinity chromatography on glutathione–agarose beads. Providing the condition that tested substance is the ligand of the receptor, the complex GST/X-receptor-ligand is formed and retained on the column. Thereafter, the complex is eluted from the column and the eluate is analyzed by western bloting with direct autoradiography detection.

2.2.3. Transfection Assays

Transfection assays are in principle simple and demonstrative method suitable for the studies on the receptors activators/antagonists. Most of all it allows to study the effect of mutations within a receptor and to explore the receptor responsive elements in DNA promoter region. Important step is the careful selection of the cell type for the individual purpose, i.e. the cells expressing/lacking certain factors.

Transient transfections are routinely used, when plasmids with cloned reporter, vector (AhR or GR) and control genes (beta-galactosidase) are transfected to the cells. Circular DNA is imposed on the cells usually by lipofection (FuGENE agent) or high calcium shock. The genes for luciferase (LUC) or chloramphenicol acetyltransferase (CAT) are fused to the promoter region and serve as reporters. Variety of factors, i.e. cell type, vector, reporter, and eventually co-transfected factors under investigation, can be combined when using this assay. These assays, however, do not reflect perfectly the physiological situation, because of over-expression of transfected genes. It is rather convenient for the direct proof of the particular factor involvement in the process.

Stably transfected cells are the model much closer to the physiological conditions as compared with transiently transfected cells. Albeit, the construction of the stably transfected cell line is experimentally difficult, once the stable transfection is accomplished, i.e. the desired clone isolated, further manipulation is such as easy as with unmodified cell line. Human cervix carcinoma cell line (HeLa) stably transfected with the glucocorticoid responsive reporter gene MMTV-Luc-SVNeo fused to the luciferase gene was described [19]. Since HeLa cells express endogenous GR, the activation of the receptor is directly monitored by luciferase activity measurement. Similarly the rat hepatoma cell line (H4IIE) stably transfected with a luciferase reporter gene under the control of dioxin-responsive enhancers was successfully used for the studies on AhR activation [20]. Again, H4IIE cells express endogenous AhR, thus luciferase activity is dependent on AhR activation.

2.2.4. Electrophoretic Mobility Shift Assay (EMSA)

The electrophoretic mobility shift (EMSA) assay is a simple and rapid method for detection and analysis of proteins-DNA interactions [21, 22]. The assay is based on the observation that complexes of protein and DNA migrate through a non-denaturing polyacrylamide gel more slowly than uncomplexed DNA fragments or double-stranded oligonucleotides. The gel shift assay is performed by incubating a purified protein, or a mixture of proteins (e.g. nuclear extract preparation), with a ^{32}P end-labeled DNA fragment containing the studied protein binding site. The reaction products are then analyzed on a non-denaturing polyacrylamide gel. The specificity of the DNA-binding protein for the putative binding site is established by competition experiments using non-labeled DNA fragments or oligonucleotides containing a binding site for the protein of interest or other unrelated DNA sequences.

2.3. Cellular Localization and Modification Studies

The methods described above give us a robust image of the events and factors participating in the receptors activation, however, this image must be complemented by further knowledge. Very important information is gained from the analysis of the receptor cellular localization and its covalent modification.

2.3.1. Nucleo-Cytosolic Shuttling of AhR and GR

The characteristic feature of AhR and GR receptors functioning is their nucleo-cytosolic shuttling, i.e. the receptors nuclear import and export. The cellular localization of the receptor is studied using the following techniques:

Nuclear extracts analysis – Nuclear and cytosolic fractions are isolated from the cells by two step extraction of proteins using the combination of “low salt buffer” and “high salt buffer”. Differential centrifugation than allows to obtain the nuclear protein extracts and cytosolic fraction [23]. The protein fractions are analyzed for the receptors presence by western blot.

Immune-histochemistry – The cells are grown on glass cover slips. Following the experiments, the cells are fixed, permeabilized and incubated sequentially with primary antibody against the receptor and secondary antibody conjugated with a label of choice. The receptor cellular localization is than monitored by microscopy.

Fluorescent proteins – Another approach for the visualization of the protein localization within the cell is the use of cells transiently transfected with chimera proteins. The studied protein (receptor) is fused to the fluorescent protein. The localization of the receptor chimera is visualized by fluorescence microscopy. Yellow fluorescent protein (YFP), green fluorescent protein (GFP) and cyan fluorescent protein (CFP) were successfully used in the studies of glucocorticoid receptor nuclear translocation [24, 25, 26].

2.3.2. Modification of the Proteins

Covalent modification of the receptor is the prominent mode of the control for its transcriptional activity and its half-life (turnover). In particular, phosphorylation and dephosphorylation of the receptors modulates their transcriptional activity. In the present, the phosphorylation status of the receptors is routinely evaluated by western blot using commercially available antibodies against the individual phosphorylated forms of the protein. Another covalent modification, which undergo both AhR and GR, is the protein ubiquitination that label the protein for proteasome mediated degradation. A convenient procedure for evaluation of the receptor ubiquitination status is the immuno-precipitation of the receptor with antibody against the receptor and subsequent western blot using detection with antibody against ubiquitine.

3. Glucocorticoid Receptor

3.1. Cellular Signaling by Steroid Receptors

Steroids are lipophilic substances destined to relay certain kind of signal to a cell. Receptors are dutiful recipients of this information translating it into expression of specific genes [4]. Specificity of the transcription carries many questions as yet unanswered: 1) how does the ligand-receptor interaction increase or decrease transcription of particular genes; 2) how is tissue specificity achieved; and 3) how is a gene up-regulated in one tissue and down-regulated in others. Steroid hormones utilize carrier proteins to reach their destination cells but in fact may be distributed equally suggesting high impact only within close vicinity of the source tissue. Physiology ascertains us otherwise: hormones work throughout our bodies essentially independent of the distance from their site of formation. Therefore other factors must affect the influence of hormones on cellular behavior.

Receptors are the frontline regulators. Omitting plasma membrane receptors, the nuclear ones are of interest of this chapter. Nuclear receptors are quite often cytoplasmic, i.e. they are present in the cytosol rather than in the nucleus (Fig. 1). GR is usually present in the unliganded form in the cytosol associated with multiple chaperone proteins [27]. This multiprotein complex may completely or in part dissociate triggering translocation of the liganded receptor to the nucleus. There it may bind DNA directly or associate with other co-factors and only then bind the target DNA sequence. The co-factors likely consitute the tissue specificity as kinases/phosphatases were implied in GR interactions. Futhermore, certain stages of the cell cycle are of importance because activity of many phosphorylating proteins is linked to these events.

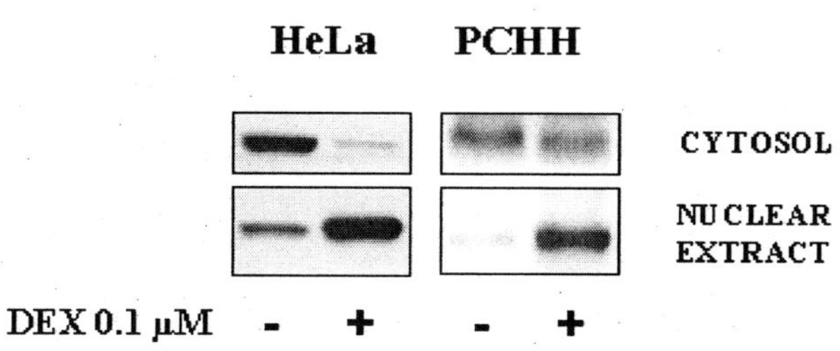

Fig.1 Hormone dependent GR nucleo-cytosolic shuttling. Human cervix carcinoma cells (HeLa) and primary cultures of human hepatocytes were treated 30 min with dexamethasone (DEX; 0.1 μM final concentration) and/or DMSO as vehicle for control. Cytosolic and nuclear extracts were isolated and the extracts were subjected to the western blot analyses. Membranes were probed with specific antibodies against human glucocorticoid receptor (hGR). The representative western blots of hGR protein are shown.

Overall four levels of steroid receptors transcriptional activity control are in force: 1) binding of a ligand, although unliganded activity was documented as well; 2) cytosol to nucleus traffic; 3) DNA binding in the presence or absence of co-factors; and 4) proteasome mediated degradation, serving as a negative feedback control.

3.2. Glucocorticoid Receptor Structure and Function

The glucocorticoid receptor (GR) belongs to the steroid/thyroid receptor family. It is an indispensable factor for cell survival and function, playing a crucial role in cell homeostasis, growth and development, cell differentiation and proliferation, the regulation of metabolic processes, host defense, inflammation combat, and many other important functions [28, 29]. Expressed in virtually all tissues, it regulates gene expression in a cell type-specific manner. The wide spectrum of genes is transcriptionally controlled by the GR, some are down-regulated, e.g. cytokines (tumor necrosis factor alpha; TNFα), enzymes (cyclooxygenase 2; COX-2), receptors (interleukine 2 receptor; IL-2R), whereas others are up-regulated, e.g. β2-adrenoreceptor, tyrosine aminotransferase, interleukine 1 receptor alpha (IL-1Rα), inhibitor kappa B alpha (IκB-α) [30]. hGR has a crucial position in hepatic drug metabolism: it controls expression of orphan receptors, PXR and CAR, and consequently the expression of phase I [31], phase II [32], and phase III enzymes [33]. Any alteration in hGR activity/expression may have potentially deleterious impact on the detoxication capacity of the organism. For instance, it was demonstrated that the expression of drug metabolizing enzymes is reduced due to GR activity suppression by cytokines [34]. Human GR exists in two forms, transcriptionally active GR-α (MW 95 kD) and the inactive GR-β (MW 90 kD), which differ at their carboxy termini [35]. The alternative translation of human GR resulting in synthesis of 94 kD GR-α and 91 kD GR-β was recently suggested [36]. GR protein constists of a ligand binding domain, a DNA binding domain containing two zinc-fingers, and two transactivation domains AF1 and AF2 [37]. The transcriptional activity of the GR depends additionally on co-activators the role of which is to facilitate recruitment of the basal transcription machinery or to remodel chromatin. The homologous coactivators CBP (cAMP-response element binding protein binding protein) and p300 interact directly with AF1 [38] while the indirect interaction of CBP/p300 with AF2 of the hGRα, is mediated by SRC1/NCoA-1 coactivator [39]. The interaction in AF2 site is mediated by `LXXLL' motifs in *NR boxes* in the coactivator protein. GR interacts with the DRIP (vitamin D3 receptor-interacting protein) complex of proteins, related to the TRAP (telomerase regulation-associated protein) and ARC (activator-recruited cofactor) complexes [40, 41]. These complexes interact with and coactivate several nuclear receptors. In the absence of a ligand (glucocorticoid; GC), GR is predominantly localized in the cytosol in complex with chaperone proteins hsp90, hsp70, and hsp90-binding immunophilins. The complex of GR and its chaperones moves along intact microtubules from the cytosol to the nucleus [42]. Regardless the presence of GC, a dynamic equilibrium exists between the cytosolic and nuclear GR [43]. This equilibrium is affected by nuclear import of GR mediated by at least two amino acid sequences [44], as well as nuclear export facilitated by DNA-binding domain of GR [45] and mediated by calreticulin [46]. Upon the hormone binding, GR translocates to the nucleus where it forms a GR/GR homodimer that binds Glucocorticoid Responsive Element (GRE) sequences in DNA and triggers transcription of target genes [47]. Recent work involving transgenic mice showed that GR dimerization is not required for all GR activities *in vivo*, though [47].

3.3. Glucocorticoid Receptor Regulation

Several modes of GR activity regulation exist comprising the following:

Hormone dependent activation: The GR is typical hormone-activated receptor, the transcriptional activity of which is strictly limited to the presence of glucocorticoid. It implies that all conditions affecting GC systemic concentration, hormone uptake by the cell, or its binding to the receptor will result in GR inhibition. In addition to the classical concept of hormone dependent activation of steroid receptors, ligand independent activation was described in case of estrogen related receptor alpha [48] and aryl hydrocarbon receptor [49]. Thus, co-activator proteins or peptides can be also considered ligands besides steroid hormones and we have to consider other means of steroid receptor activation as viable alternative *in vivo* [50].

Phosphorylation: The GR is a phosphoprotein [51] that is additionally phosphorylated during the ligand binding process [52]. Four major GR phosphorylation sites with an impact on receptor activity have been identified [53]. Phosphorylation of rat GR at Ser224 and Ser232 by cyclin-dependent protein kinases (CDKs) leads to rat GR activation, whereas phosphorylation at Thr171 and Ser246 by mitogen-activated protein kinases (MAPKs), has an inhibitory effect on rat GR [53]. Two MAPKs, c-jun N-terminal kinase (JNK) and extracellular signal-regulated kinase (ERK), negatively modulate rat GR activity by phosphorylation at Ser246 [54]. Consequently, any factors (e.g. various chemicals, mitogens, cytokines) activating/inhibiting CDKs or MAPKs will modulate GR activity.

Interaction with transcription factors: Cross-talk between the GR and other transcription factors (TFs) influence GR transcriptional activity. Comprehensive information is given in several review articles [30, 55]. Nuclear factor kappa B (NFκB) and activator protein 1 (AP-1) are the most important TFs affecting GR activity. Both factors are activated via multiple pathways, which modulate transcription of inflammatory genes. The binding of cytokines, growth factors or chemokines to their respective receptors leads to the activation of signal transduction pathways, including MAPKs, resulting in NFκB and AP-1 activation [30]. In its inactive form, NFκB, a heterodimer of p50 and p65 proteins, is localized in the cytoplasm in complex with its inhibitor IκB-α. Upon stimulation, IκB-α is degraded and p50/p65 translocates to the nucleus where it binds DNA and triggers the transcription of target genes [56]. Similarly, in unstimulated cells, AP-1 is sequestered in the cytoplasm with scaffold protein, and when activated the heterodimer c-fos/c-jun and/or homodimer c-jun/c-jun translocates to the nucleus where it regulates transcription [57]. Interaction between GR and NFκB and/or AP-1 occurs at different levels, resulting in orchestrated inhibitory machinery [30, 55, 58]. One mode of the interaction between GR and TFs is the functional transcriptional combat when GR up-regulates genes involved in anti-inflammatory response and down-regulates genes involved in inflammatory response. It is an opposite of transcriptional activity of NFκB and AP-1. There is established the dynamic equilibrium between transcriptionally active GR and TFs, and the fluctuations in cytokines or steroid hormone concentrations determine the overall biological response. The example of the regulatory feedback is the transcriptional activation of NFκB inhibitor IκB-α by glucocorticoids [59, 60]. Another mode of GR activity modulation by TFs is the protein-protein interaction. Indeed, direct protein-protein interaction between GR protein and NFkB and/or AP-1 constituents has been described, occurring both in cytoplasm and the nucleus, resulting in physical interference with GR protein transcriptional activity [60, 62, 63]. In addition, competition for essential coactivator proteins, e.g. CBP, p300, and SRC-1, may be the cause for AP-1 effects on GR transcriptional activity [64].

Receptor degradation: Further mode of GR activity regulation is by the reduction of GR half-life by GCs. Although considered to be a constitutively expressed gene, there are reports on the auto regulation of human GR mRNA expression by GCs [65]. In the presence of GCs the half-life of GR protein drops to 30 - 50 % of its initial value [66]. Moreover, it was reported that the synthetic GC dexamethasone (DEX) causes mouse GR protein degradation mediated by the proteasome-ubiquitin pathway [67].

4. Aryl Hydrocarbon Receptor

4.1. Signal Transduction by Xenoreceptors

The presence of hormones and xenobiotics alike must be dealt with properly for which reason Nature developed appropriate receptors. Given this role for xenoreceptors they serve as relays positioned between genome and the extracellular signals to which cells must respond in order to survive. Role of these regulatory proteins in pathological processes is therefore not surprising. In fact, any molecule coming from outside, be it a hormone or a foreign substance, is an extracellular stimulus. Consequently, both kinds of receptors share many common features, among them direct or indirect involvement in inflammation process.

The main objective of xenoreceptors is to modulate the expression of proteins involved in metabolism of xenobiotics potentially dangerous to the cell. For example, nuclear receptors PXR (pregnane X receptor, NR1I3) and CAR (constitutive androstane recpetor, NR1I2) control the expression of distinct but overlapping arrays of genes including the CYP2/3 families. The crosstalk between both PXR and CAR and other nuclear receptors allows for the establishing and prediction of functional interferences between detoxification and other signaling and metabolic pathways [1]. The tangle of nuclear receptor network is complex and is continually unveiled [3]. In this sea of receptors sticks out aryl hydrocarbon receptor (AhR) as a sore thumb constituting a standalone xenoreceptor insubordinate to other nuclear receptors.

4.2. Aryl Hydrocarbon Receptor Structure and Function

The aryl hydrocarbon receptor (AhR), 848 a.a. protein with a theoretical mass of 96,147 Da [68, 69], belongs to the PAS (Per-ARNT-Sim) family of transcription factors. PAS family is a gene family of a structurally related transcription factors with characteristic structural motifs designed as basic helix–loop–helix (bHLH). The bHLH motif, localized in the N-terminal region, consists of two amphipathic α-helices separated by a non-helical loop. The HLH and the PAS regions mediate AhR/ARNT heterodimer formation, while the basic region is required for binding to DNA. The ligand binding domain (LBD) of AhR is located in the sequence of 230–431 a.a. from the N-terminus, partly overlapping with the binding site for hsp90 that keeps AhR structurally competent to bind ligands [70].

Analogically as in case of GR, the unliganded AhR resides in the cytosol in a multiprotein complex with two hsp90, one p23 and an immunophilin-like protein termed XAP, AIP or ARA9, the latter being involved in the correct folding and stabilization of AhR. Hsp90 chaperone prevents transcriptional activation of the AhR but keeps the receptor in a conformation that facilitates ligand binding. Upon ligand binding, the chaperon proteins dissociate, and AhR translocates to the nucleus, where it forms a 180-220 kDa heterodimer with aryl hydrocarbon receptor nuclear translocator (ARNT) [71, 72]. This heterodimer binds to the specific DNA region termed dioxin or xenobiotic response element (DRE or XRE), which has a core sequence 5'-TNGCGTG- 3' [70], and triggers gene expression.

AhR has been considered an orphan receptor for a long time, however, the existence of an endogenous ligand was indirectly supported by numerous observations [73]. As a candidates for the endogenous ligands are considered in particular indole derivatives originating from intermediary metabolism [74]. The best evidence for its existence came from the studies on AhR null mice, the growth rates of which were retarded as compared to wild type mice for the first 3 weeks of life. The mutant mice were defective in liver and immune system development, and several phenotypic defects have been reported for adult AhR-null mice, such as retinoid accumulation in liver, abnormal hepatic and kidney vascular structures, decreased liver size and hepatic fibrosis [70]. Exogenous activation of AhR, e.g. by exposure to dioxin or polyaromated hydrocarbons, results in a number of toxic effects, including cancer, immunosuppression, and endocrine disruption [75]. Thus, the role of AhR receptor in living organisms is ambiguous, resembling "*Dr. Jekyll and Mr. Hyde*" story. On one hand, AhR exogenous activation results in a number of toxic effects, on the other hand, sustained activation of AhR by endogenous ligands is essential for the correct development and function of living organisms. AhR plays an important role in the regulation of drug metabolizing enzymes, when it controls the expression of cytochromes P450 CYP1A1, CYP1A2 and CYP1B1, as well as several phase II metabolizing enzymes, such as NAD(P)H: quinone-oxidoreductase, glutathione-S-transferase (GST), and uridine 5'-diphospho-glucuronosyltransferase (UGT) [68, 70].

4.3. Aryl Hydrocarbon Receptor Regulation

Similar to the activity of GR and other nuclear receptors, several modes of AhR activity regulation exist:

Ligand dependent activation: This is the classical means of nuclear receptors activity regulation, involving activation by both exogenous and endogenous ligands. Classical exogenous AhR ligands such as dioxin (TCDD), 3-methylcholanthrene (3-MC), benzo/a/pyrene (BP) or α- naphthoflavone [76] share several common structural features: they are hydrophobic, planar or co-planar molecules of polycyclic structure. Nevertheless, several non-polycyclic and non-planar compounds like omeprazole, thiabendazole or carotenoids including canthaxanthin activate the CYP1A1 gene as well. However, they are not or very weak AhR ligands as compared to those mentioned above [76]. The greatest exposure to AhR ligands comes from food (e.g. flavonoids, carotenoids, phenolics), however, the majority of these chemicals are relatively weak AhR ligands. Flavonoids flavone and diosmin act as AhR agonists and induce CYP1A1, CYP1A2 and CYP1B1 activities [77].

Receptor degradation: The proteolytic degradation of transcription factors is an established mechanism for regulating signal transduction pathways. The AhR protein is degraded upon exposure to the ligand and nuclear export both *in vivo* and *in vitro*. The magnitude of degradation differs among various cell types and it is proteasome-ubiquitin dependent. Following the exposure of cells to TCDD, the concentration of AhR in nucleus reaches the maximum in 1–2 h, and then rapidly drops, as AhR is degraded [78]. When complexed with hsp90 and immunophilin-like proteins, AhR is inactive and these proteins participate in masking the nuclear localization signal (NLS), resulting in the cytoplasmic retention of AhR [70, 73]. Obviously, AhR contains a nuclear export signal (NES), which is important for AhR removal from the nucleus [70].

Protein-protein interactions: As NFκB and AP-1 restrict transcriptional activity of GR, so several proteins/receptors modulate transcriptional activity of Ah receptor. A mouse protein closely related to the AhR, designated as **aryl hydrocarbon receptor repressor** (AhRR), has been identified. It is localized in the nucleus and does not have capacity to bind TCDD. AhRR interacts with ARNT but AhRR/ARNT heterodimers are transcriptionally inactive. The expression of AhRR is induced by AhR, thus serving as negative feedback loop [70]. The important player modulating AhR activity reciprocally is **estrogen receptor** (ER) and the existence of cross-talk between AhR and ER was hypothesised. For instance, estrogen-dependent hepatocellular carcinogenesis induced by TCDD was observed in female but not male rats [72]. Moreover, decreased formation of estrogen dependent uterine and mammary tumors in rats fed by TCDD in the diet suggested that TCDD blocked ER action [72]. The functional interplay between AhR and retinoid acid receptors (RARs) signaling pathways, vitamin D receptor (VDR), and peroxisome proliferator-activated receptors (PPARs) have recently been described [76, 79, 80].

Phosphorylation: AhR ligands stimulate signaling cascades similar to those initiated on the cell surface by growth factors, hormones, neurotransmitters and extracellular signals, and share with classical mitogenic factors the propagation of the signal including elevation of intracellular Ca^{2+} and cAMP levels, and the activation of protein kinases (PKs) and protein phosphatases (PPs). AhR/ARNT is probably regulated via phosphorylation, because the binding activity of the AhR/ARNT heterodimer to the XRE is abolished by the phosphatase treatment [81]. TCDD causes an immediate increase in protein kinase C (PKC) and tyrosine kinase activities in hepatocyte membranes. Evidence for PKC involvement in AhR activation comes from the finding that specific inhibition of PKC by staurosporine [76] prevents the ligand-induced DNA-binding of AhR/ARNT heterodimers and leads to suppression of CYP1 expression. Phosphorylation of AhR in vitro has been localized to the C-terminal part of the peptide [81]. AhR and ARNT are both phosphorylated on threonine residues [81], suggesting that a serine/threonine kinase may be involved in the regulation of these proteins.

5. Microtubular Network in AhR and GR Signaling

5.1. Microtubules the Vital Network of the Cell

The primary types of structures constituting the cytoskeleton are microfilaments, microtubules, and intermediate filaments. Microtubules are the hollow cylinders, 20-25 nm in diameter. A heterodimer consisting of α– and β-tubulin is the basic structural unit of microtubules, the α– and β–tubulins being homologous but not identical proteins. Microtubules act as a dynamic scaffolding to determine cell shape and provide a set of "tracks" for cell organelles and vesicles to move on. Microtubules also form the spindle fibers for separating chromosomes during mitosis; when arranged in geometric patterns inside flagella and cilia, they are used for locomotion. α-tubulin has a bound molecule of GTP that is not hydrolyzed, while β-tubulin may bind either GTP or GDP. Under certain conditions β–tubulin can hydrolyze its bound GTP to GDP plus P_i, release the P_i, and exchange bound GDP for GTP. Thus, microtubules possess intrinsic GTPase activity, offering the opportunity to interfere with GTP/GDP dependent processes. Microtubules function is closely interconnected with microtubule associated proteins (MAPs). They are tissue and cell type specific, high molecular weight proteins (200-300 kDa) or the tau proteins (20-60 kDa). One domain binds to tubulin polymers or unpolymerized tubulin, which speeds up polymerization, facilitates assembly and stabilizes the microtubules. Furthermore, some MAPs may bind to special sites on the α–tubulin that form after it is in the microtubule. These sites are important marker sites for stabilized microtubules, because they disappear when microtubules are depolymerized. The MAPs include kinesins and dynein which "walk" along the microtubules in opposite directions. The kinesins move the vesicle along towards the plus end and dynein walks towards the minus end.

As will be discussed, several chemicals may affect microtubules assembly, stability or function. Colchicine, demecolcine, and nocodazole inhibit polymerization by binding to tubulin and preventing its addition to the plus ends. Vinblastine and vincristine aggregate tubulin and lead to microtubule depolymerization. Taxol stabilizes microtubules by binding to a polymer.

Surprisingly, there is little evidence for the role of cytoskeleton and/or the microtubules network in AhR and GR function and regulation. Moreover, data published on the importance of MTs in GR function are somewhat controversial.

5.2. Role of Microtubule Network in GR and AhR Receptors Signaling

Analyzing microtubules necessity for GR functioning, we consider ligand binding, GR-protein complex dissociation, nuclear translocation, DNA binding and gene transcription. Some researchers concluded there is no involvement of microtubules in GR signaling. It was found that colchicine and demecolcine do not affect hormone dependent translocation of the GR in rat thymocytes and Ehrlich ascites cells, hence microtubules network is not involved in GR signaling [82]. Similarly, no effect of colchicine and nocodazole was demonstrated on dexamethasone induced GR nuclear translocation in COS-1 and CV1 cells transiently transfected with green fluorescent protein (GFP) – GR chimera [43]. Colchicine had no effect as well on GR mediated gene induction in human cervix carcinoma cells (HeLa) transiently transfected with GRE-CAT (glucocorticoid responsive element), a model employing endogenous GR [83]. In experiments on mouse fibroblasts it was shown that nuclear translocation of hsp90 protein, a GR partner in cytosol, is non- microtubules and non-microfilament dependent [84].

Second array of papers describes a certain role of microtubules in GR signaling, but not on hormone-dependent receptor translocation and transcriptional activity. For instance, inhibition of hsp90 dependent GR translocation by geldanamycin requires intact cytoskeleton [85]. Geldanamycin, a benzoquinone compound, binds to the ATP-binding site of hsp90 and inhibits p23 binding to hsp90 [86]. Mouse fibroblasts NIH/3T3 transiently transfected with GFP-GR and treated with demecolcine prior to treatment with geldanamycin did not display altered GR translocation. The authors explain, that intact microtubules are required for geldanamycin-mediated inhibition of GR translocation, but are not necessary for hormone-dependent GR translocation [85]. In this case, GR reaches the nucleus passively by diffusion, which is a slower process not inhibited by geldanamycin. Also, effective inhibition of GR nucleo-cytoplasmic shuttling by okadaic acid in the same *in vitro* model requires intact microtubules [87]. Inhibition of GR shuttling by okadaic acid (9,10-Deepithio-9,10-didehydroacanthifolicin), an inhibitor of serine/threonine phosphatases, was reversed by colcemid. Co-transfection of NIH/3T3 cells with the peptidylprolyl isomerase domain of hsp90-binding immunophilin FKBP52 inhibited hormone dependent GFP-GR nuclear translocation. This inhibition was abrogated by demecolcine-induced microtubules disruption and was restored after demecolcine withdrawal [88]. Taken together, it seems that microtubules play a role in proper GR signaling despite the fact that in cells with disrupted microtubules hormone induced GR translocation is maintained to some extent. Comprehensive review on GR movement through cellular compartments gives the notion that indeed a supporting "grapevine" is necessary for GR activity [27]. We have recently demonstrated that microtubules disruption in primary human hepatocytes by colchicine and nocodazole perturbs hGR dependent pregnane X receptor (PXR)/constitutive androstane receptor (CAR)-cytochrome P450 cascade by a mechanism involving inhibition of hGR translocation to the nucleus [26]. We have also observed inhibition of dexamethasone dependent hGR-GFP chimera nuclear translocation by colchicine in transiently transfected human embryonic kidney (HEK293) cells. On the other hand, there was no effect of COL on luciferase activity in transiently transfected (hGR/GRE-LUC) human hepatoma cell line

HepG2, while in human cervical carcinoma cell line HeLa stably transfected with GRE-LUC reporter we found inhibition of hGR activity by colchicine and nocodazole [26].

Another mode how microtubules could influence GR is the interplay between GR and discrete transcription factor. For instance, microtubule interfering agents were found to activate c-jun N-terminal kinase (JNK) and stress-activated protein kinase (SAPK) in human breast cancer cells (MCF-7) and ovarian cancer cells (BR) [89]. The implication of JNK in inhibition of GR transcriptional activity was clearly described in HeLa cells and osteosarcoma cells ectopically expressing GR (SAOS2) [54]. Furthermore, enhancement of GR nuclear export by JNK mediated phosphorylation was reported in COS-7 cells [90]. The differential and selective role of MAPKs and CDKs in GR transcriptional activation has also been published [53]. We demonstrated recently a modus operandi affecting GR activity: microtubules disarray affects GR protein stability by causing proteasome-ubiquitin-mediated ultra-rapid dose- and time-dependent GR degradation [91]. Figure 2 shows the microtubular network collaps by colchicine in HeLa cells. GR stability has been evaluated in the past but the connection with intact microtubules was not made [67].

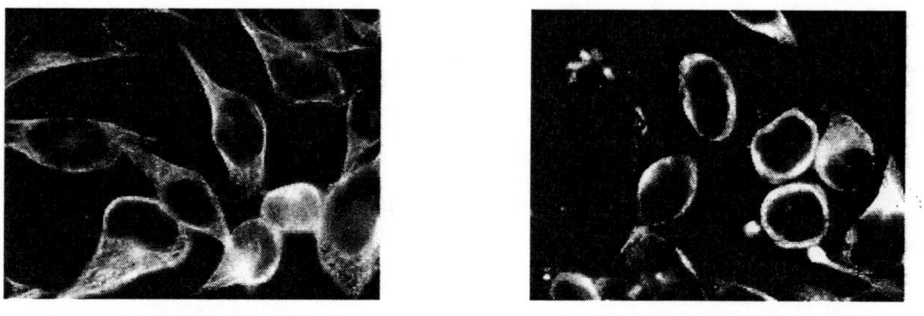

CONTROL COLCHICINE

Fig.2. Disruption of microtubules by colchicine in HeLa cells. HeLa cells were treated with 1 □M colchicine or DMSO as vehicle for 30 minutes. Microtubular network was then visualized by immunohistochemical detection and fluorescent microscopy imaging

One of the reasons for divergent data is the use of overexpression systems for studying involvement of MTs in GR signaling. It appears that transient transfection may be performed in any cell type regardless of the presence of endogenous GR. However, the actual equipment of the cell consisting of auxiliary proteins should be taken into account, in particular the stoichiometry with chaperones, kinases/phosphatases, transcription activators and/or repressors. All of these, including microtubules, ought to be considered the binding site possibly saturated by unusual amount of GR present during overexpression. Although cell lines, the tool for transient transfection, cannot be considered exactly physiological, overexpression of a certain protein is bound to create "unphysiological" state in any case. Therefore, behavior of GR in this system must be interpreted very carefully.

On the other hand, there is limited information only on the role of microtubules in aryl hydrocarbon receptor signaling. The data available reveal about the association between AhR transcriptional activity and the synchronization of the cell in M-phase of the cell cycle by microtubules disruption [92, 93, 94, 95].

5.3. GR and AhR Activity and Cell Cycle

If GR transcriptional activity was limited by nuclear trafficking which depends on the microtubule network, and some laboratories suggest that this is so [26, 85], then GR-dependent transcription would cease at certain points of naturally occurring cell cycle, namely late G2, M, and early G1 phases. GR-dependent transcription is repressed during the M phase only and this is proposed to be due to chromatin condensation rather than impaired nuclear shuttling [96]. In the case of AhR, it interacts with the retinoblastoma protein (RB1), and the AhR– RB1 complex hinders progression from the G1 to the S phase of the cell cycle by blocking E2F transcription factor-mediated expression of S phase genes [68]. Thus, activated AhR obviously participates in cell cycle arrest at the G1/S boundary. The level of CYP1As in these cell types is critical, because CYP1As degrade AhR ligands and thus prevent the block in progression in G1/S. Therefore it appears the AhR affects the cell cycle rather than the opposite as in the GR scenario. On the other hand, AhR activity ceased during synchronization of cells in G2/M phase by nocodazole treatment [95], the effect is somewhat enhanced by using colchicine (Dvořák, unpublished results).

In fact cell tends to limit the activity of receptors during specific phases of the cell cycle in order to "economize" resources. One of the reasons is the shift in receptor/DNA binding site ratio, because during mitosis the number of binding sites effectively doubles. Wide array of mechanisms is at the cell´s disposal, e.g. complexation, phosphorylation, degradation, to achieve the sovereignty within its realm over receptors in time of proliferation.

Glucocorticoids are frequently used to modulate cell proliferation but a mutual competition likely exists between GR activity and cell cycle progression because proteins associated with cell cycling inhibit GR activity. Protein 14-3-3s, which plays critical role in cell cycle arrest/progression, has been shown to interact with GR and negatively suppress transcriptional activity of the receptor [97]. Important appears the cross-talk of GR with $p27^{Kip1}$, a protein expressed at high levels in differentiated, non-cycling cells [98]. Hepatocytes, which are non-proliferating cells mostly in the quiescent G0 state, suffer the loss of GR activity as the consequence of microtubules disarray. Is it possible that following microtubule interfering agent treatment hepatocytes enter the cell cycle *in vivo* and are subsequently arrested in a cycle phase, likely the M phase? Hepatocytes may be primed, i.e. induced to enter the cell cycle, when four transcription factors are activated: NFkB, AP-1, C/EBP beta, and STAT3 [99]. The last factor is activated by IL-6 which was shown to affect GR-dependent transcription in primary human hepatocytes [100]. Systemic administration of IL-6 *in vivo* causes liver growth and hence may function as a direct hepatic mitogen [101]. The two situations are not identical because growth factors, which are required for the cell cycle to complete, were absent in the primary hepatocytes but very likely present in the *in vivo* experiment. Perhaps the transition point between cell cycle phases, or a cell status which mimics such a transition, is the key feature in GR transcriptional activity suppression.

Treatment with microtubule interfering agents is one of the "non-physiological" situations under which cell cycle is prolonged or even stalled causing deleterious effects on GR and likely AhR activity. Considering the case of IL-6, it may be triggering and at the same time stalling the G0 - G1 phase transition, thus causing the observed phenomenon of GR repression [100]. On the other hand, microtubule interfering agents may be mimicking M phase arrest resulting in GR repression as well [26, 102]. One is tempted to speculate that microtubules constituting the cytoskeleton and microtubules present in the nucleus are linked and integrity of this complex system helps the GR to interpret cell cycle transitions with M phase being the major change.

If AhR activity follows, to a certain extent, analogy with GR then it ought to be affected by cell cycle as well. Data published on CYP 1A expression support this hypothesis of subdued AhR activity during M phase [93, 95]. We found that IL-6 had an inhibitory effect on TCDD-induced CYP 1A1 expression in human hepatoma cells (Dvořák, unpublished data). While this data alone are largely inconclusive in terms of cell cycle, HepG2 cells are proliferating and certain unknown number of the cells is in M phase at any given time, it suggests that IL-6 fills the role of a mitogen increasing the number of cells in M phase and in consequence lowering AhR activity.

5.4. Regulatory Cross-Talk between GR and AhR

Several lines of evidence indicate the existence of a regulatory cross-talk between AhR and GR receptors. For instance, CYP1A induction by dioxin was augmented by synthetic glucocorticoid dexamethason and by GR agonists, cortisol and prednisone, in Poeciliopsis lucida hepatocellular carcinoma cell line. The involvement of GR in this process was confirmed, as the addition of GR antagonists PCN (pregnenolone-16alpha-carbonitrile) or RU486 reduced the potentiation of CYP1A induction [103]. Similarly, the study on H4IIE rat hepatoma cells demonstrated that TCDD-dependent CYP1A1 induction is augmented or suppressed by dexamethason or 17beta-estradiol (E2), respectively [104]. Dexamethason-mediated potentiation was found to depend on posttranscriptional processing, while E2-mediated suppression did not rely on the synthesis of protein factors. Presumably it might hinder the formation of the activated TCDD/AhR complex and so the subsequent binding on DRE [104]. Furthermore, the inhibition of GR-mediated transcription and chromatin remodeling by estrogen agonists, but not antagonists in MCF-7 cells was reported, when enhanced GR degradation via the proteasome pathway was observed in the presence of estrogen agonists [105]. *In vivo* study described synergistic induction of cleft palate in mice embryos when co-treated with TCDD and hydrocortisone [106]. The effect was ascribed to the interactive regulation of AhR and GR, where the presence of glucocorticoid reduced the levels of GR protein and mRNA, but enhanced the amount of AhR. Reciprocally, TCDD down-regulated AhR protein and mRNA, but in parallel augmented GR levels [106]. As described earlier (Section 4.3.), AhR activity is reciprocally modulated by estrogen receptor ER, and the existence of cross-talk between AhR and ER is well known [49, 72]. Collectively, it seems that regulatory "triangle cross-talk" between AhR, ER and GR receptors exists (Fig. 3). The activity of AhR is positively modulated by GR and vice versa, while reciprocal negative modulation between AhR and ER occurs. Finally, ER negatively modulates GR activity, whereas GR effect on ER was not reported yet.

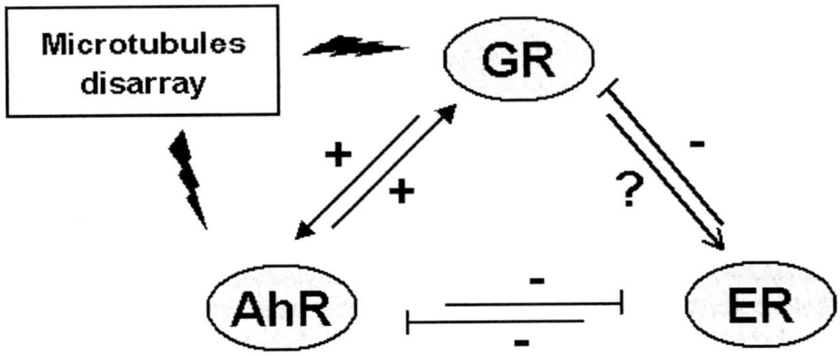

Fig.3. Interactive regulatory triangle between AhR, GR and ER receptors. The activity of aryl hydrocarbon receptor (AhR) is positively modulated by glucocorticoid receptor (GR) and vice versa. AhR and estrogen receptor (ER) activities are reciprocally negatively modulated. ER negatively modulates GR activity, whereas GR effect on ER is not known yet

6. Conclusion

In physiological systems with disordered microtubules networks, a clear inhibition of GR by various mechanisms was observed; protein-phosphatases/kinases recruitment, nucleo-cytoplasmic shuttling inhibition, and decrease in transcriptional activity. By contrast, in models employing GR overexpression, transient transfection models in particular, GR function appeared to be hardly affected. The same may be true, and possibly already unraveled, in case of AhR. It is then possible that both receptors are maintained in the cytoplasm through a specific cytoplasmic retention signal and that the transfer of either receptor reflects the saturation of cytoplasmic retention sites. In our view, microtubules, their dynamic equilibrium, and likely the link between cytoskeleton and nuclear microtubules, constitute these retention sites. Proliferation status, transformation, transfection status, and interspecies variability all contribute to the technical and experimental limitations for research reaching for the conclusive answer of microtubular network involvement in GR and AhR activity. GR and AhR entanglement in the essential cellular processes, as well as their specific assignments in the xenobiotic biotranformation enzymes expression, appoints them the pillars of cellular responsiveness to outside stimuli. Microtubules are the matrix necessary to support proper functioning of the system and providing a link to the cell´s life cycle.

References

[1] Pascussi, JM; Dvorak, Z; Gerbal-Chaloin, S; Assenat, E; Drocourt, L; Maurel, P; Vilarem, MJ. Regulation of xenobiotic detoxification by PXR, CAR, GR, VDR and SHP receptors: Consequences in physiology. In: Gossen, M; Kaufmann, J; Triezenberg, SJ editors. *Handbook of Experimental Pharmacology*: Vol 166 Transcription factors). Berlin: Springer-Verlag; 2004; pp 409-435.

[2] Pascussi, J.M.; Busson-Le Coniat, M.; Maurel, P.; & Vilarem M.J. (2003) Transcriptional analysis of the orphan nuclear receptor constitutive androstane receptor (NR1I3) gene promotor: identification of a distal glucocorticoid response element. *Molecular Endocrinology 17*, 42-55.

[3] Pascussi, J.M.; Gerbal-Chaloin, S.; Drocourt, L.; Assenat, E.; Larrey, D.; Pichard-Garcia, L.; Vilarem, M.J. & Maurel P. (2004) Cross-talk between xenobiotic detoxication and other signalling pathways: clinical and toxicological consequences. *Xenobiotica 34*, 633-664.

[4] Helmreich, EJM. Biochemistry of cell signalling. Oxford University Press 2001, pp 190-209.

[5] Dvořák, Z; Pascussi, JM and Modrianský, M. (2003) Approaches to messenger RNA detection – comparison of methods. *Biomedical Papers Med Fac Univ Palacky Olomouc Czech Repub 147*, 131-135.

[6] Brown, T. Analysis of RNA by Northern and Slot Blot Hybridization. In: Ausubel, EM; Brent, R; Kingston, RE; Moore, DD; Siedman, JG; Smith, JA; Struhl, K editors. *Current Protocols in Molecular Biology.* New York, NY: John Wiley and Sons; 1993; pp 4.9.1.–4.9.14.

[7] Gilman M. Ribonuclease protection assay. In: Ausubel, EM; Brent, R; Kingston, RE; Moore, DD; Siedman, JG; Smith, JA; Struhl, K editors. *Current Protocols in Molecular Biology.* New York, NY: John Wiley and Sons; 1993; pp 4.7.1.–4.7.8.

[8] Rottman, J.B. (2002) The ribonuclease protection assay: a powerful tool for the veterinary pathologist. *Veterinary Pathology 39*, 2–9.

[9] Melton, D.A.; Krieg, P.A.; Rebagliati, M.R.; Maniatis, T.; Zinn, K. and Green, M.R. (1984) Efficient in vitro synthesis of biologically active RNA and RNA hybridization probes from plasmids containing a bacteriophage SP6 promoter. *Nucleic Acids Research 12*, 7035–7056.

[10] Wilkening, S. and Bader, A. (2003) Influence of culture time on the expression of drug-metabolizing enzymes in primary human hepatocytes and hepatoma cell line HepG2. *Journal of Biochemical and Molecular Toxicology 17*, 207-213.

[11] Wilkening, S.; Stahl, F. and Bader A. (2003) Comparison of primary human hepatocytes and hepatoma cell line HepG2 with regard to their biotransformation properties. *Drug Metabolism and Disposition 31*, 1035-1042.

[12] Erlich, HA. PCR *Technology: Principles and Applications for DANN amplification.* London: Stockton Press 1989.

[13] Laemmli, U.K. (1970) Cleavage of structural proteins during the assembly of the head of bacteriophage T4. *Nature 227*, 680–685.

[14] Towbin, H.; Staehelin, T. and Gordon, J. (1979) Electrophoretic transfer of proteins from polyacrylamide gels to nitrocellulose sheets: procedure and some applications. *Proceedings of the National Academy of Sciences of the United States of America 76*, 4350–4354.

[15] Bohn, M.C.; O'Banion, M.K.; Young, D.A.; Giuliano, R.; Hussain, S.; Dean, D.O. and Cunningham, L.A. (1994) In vitro studies of glucocorticoid effects on neurons and astrocytes. *Annals of the New York Academy of Sciences 746*, 243-259.

[16] Herr, A.S.; Wochnik, G.M.; Rosenhagen, M.C.; Holsboer, F. and Rein, T. (2000) Rifampicin is not an activator of glucocorticoid receptor. *Molecular Pharmacology 57*, 732-737.

[17] Nishikawa, J.; Saito, K.; Goto, J.; Dakeyama, F.; Matsuo, M. and Nishihara, T. (1999) New screening methods for chemicals with hormonal activities using interaction of nuclear hormone receptor with coactivator. *Toxicology and Applied Pharmacology 154*, 76-83.

[18] Rushing, S.R. and Denison, M.S. (2002) The silencing mediator of retinoic acid and thyroid hormone receptors can interact with the aryl hydrocarbon (Ah) receptor but fails to repress Ah receptor-dependent gene expression. *Archives of Biochemistry and Biophysics 403*, 189-201.

[19] Balaguer, P.; Francois, F.; Comunale, F.; Fenet, H.; Boussioux, A.M.; Pons, M.; Nicolas, J.C. and Casellas, C. (1999) Reporter cell lines to study the estrogenic effects of xenoestrogens. *The Science of the Total Environment 233*, 47-56.

[20] Murk, A.J.; Leonards, P.E.G.; Bulder, A.S.; Jonas, A.S.; Rozemeijer, M.J.C.; Denison, M.S.; Koeman, J.H. and Brouwer, A. (1997) The CALUX (chemical-activated luciferase expression) assay adopted and validated for measuring TCDD equivalents in blood plasma. *Environmental Toxicology and Chemistry 16*, 1583-1589.

[21] Anonymous. http://www.promega.com

[22] Fried, MG and Garner, MM. The Electrophoretic Mobility Shift Assay (EMSA) for Detection and Analysis of Protein-DNA Interactions. In: *Tietz, D editor. Nucleic Acid Electrophoresis*. Berlin-Heidelberg-New York: Springer-Verlag, 1998.

[23] Carcamo, J.M.; Pedraza, A.; Borquez-Ojeda, O. and Golde, D.W. (2002) Vitamin C suppresses TNF alpha-induced NF kappa B activation by inhibiting I kappa B alpha phosphorylation. *Biochemistry 41*, 12995-13002.

[24] Nishi, M.; Ogawa, H.; Ito, T.; Matsuda, K.I. and Kawata, M. (2001) Dynamic changes in subcellular localization of mineralocorticoid receptor in living cells: in comparison with glucocorticoid receptor using dual-color labeling with green fluorescent protein spectral variants. *Molecular Endocrinology 15*, 1077-1092.

[25] Carey, K.L.; Richards, S.A.; Lounsbury, K.M. and Macara, I.G. (1996) Evidence using a green fluorescent protein-glucocorticoid receptor chimera that the Ran/TC4 GTPase mediates an essential function independent of nuclear protein import. *The Journal of Cell Biology 133*, 985-996.

[26] Dvorak, Z.; Modriansky, M.; Pichard-Garcia, L.; Balaguer, P.; Vilarem, M.J.; Ulrichova, J.; Maurel, P. and Pascussi, J.M. (2003) Colchicine down-regulates cytochrome P450 2B6, 2C8, 2C9, and 3A4 in human hepatocytes by affecting their glucocorticoid receptor-mediated regulation. *Molecular Pharmacology 64*, 160-169.

[27] DeFranco, D.B. (2002) Navigating steroid hormone receptors through the nuclear compartment. *Molecular Endocrinology 16*,1449-1455.

[28] Schaaf M.J., and Cidlowski, J.A. (2002) Molecular mechanisms of glucocorticoid action and resistance. *The Journal of Steroid Biochemistry and Molecular Biology 83*, 37-48.

[29] Yamamoto, K.R. (1985) Steroid receptor regulated transcription of specific genes and gene networks. *Annual Review of Genetics 19*, 209-252.

[30] I.M., and Caramori, G. (2001) Cross-talk between pro-inflammatory transcription factors and glucocorticoids. *Immunology and Cell Biology 79*, 376-384.

[31] Pascussi, J.M.; Gerbal-Chaloin, S.; Drocourt, L.; Maurel, P. and Vilarem, M.J. (2003) The expression of CYP2B6, CYP2C9 and CYP3A4 genes: a tangle of networks of nuclear and steroid receptors. *Biochimica et Biophysica Acta 1619*, 243-253.

[32] Rushmore, T.H. and Kong, A.N. (2002) Pharmacogenomics, regulation and signaling pathways of phase I and II drug metabolizing enzymes. *Current Drug Metabolism 3*, 481-490.

[33] Tirona, R.G.; Leake, B.F.; Wolkoff, A.W. and Kim, R.B. (2003) Human organic anion transporting polypeptide-C (SLC21A6) is a major determinant of rifampin-mediated pregnane X receptor activation. *The Journal of Pharmacology and Experimental Therapeutics 304*, 223-228.

[34] Pascussi, J.M.; Gerbal-Chaloin, S.; Pichard-Garcia, L.; Daujat, M.; Fabre, J.M.; Maurel, P. and Vilarem, M.J. (2000) Interleukin-6 negatively regulates the expression of pregnane X receptor and constitutively activated receptor in primary human hepatocytes. *Biochemical and Biophysical Research Communications 274*, 707-713.

[35] Hollenberg, S.M.; Weinberger, C.; Ong, E.S.; Cerelli, G.; Oro, A.; Lebo, R.; Thompson, E.B.; Rosenfeld, M.G. and Evans, R.M. (1985) Primary structure and expression of a functional human glucocorticoid receptor cDNA. *Nature 318*, 635-641.

[36] Yudt, M.R. and Cidlowski, J.A. (2001) Molecular identification and characterization of a and b forms of the glucocorticoid receptor. *Molecular Endocrinology 15*, 1093-1103

[37] Beato, M.; Herrlich, P. and Schutz, G. (1995) Steroid hormone receptors: many actors in search of plot. *Cell 83*, 851-857.

[38] Chakravarti, D.; LaMorte, V.J.; Nelson, M.C.; Nakajima, T.; Schulman, I.G.; Juguilon, H.; Montminy, M. and Evans, R.M. (1996) Role of CBP/P300 in nuclear receptor signalling. *Nature 383*, 99–103.

[39] Yao, T.P.; Ku, G.; Zhou, N.; Scully, R. and Livingston, D.M. (1996) The nuclear hormone receptor coactivator SRC-1 is a specific target of p300. *Proceedings of the National Academy of Sciences of the United States of America 93*, 10626–10631.

[40] Ito, M.; Yuan, C.X.; Malik, S.; Gu, W.; Fondell, J.D.; Yamamura, S.; Fu, Z.Y.; Zhang, X.; Qin, J. and Roeder, R.G. (1999) Identity between TRAP and SMCC complexes indicates novel pathways for the function of nuclear receptors and diverse mammalian activators. *Molecular Cell 3*, 361–370.

[41] Rachez, C.; Lemon, B.D.; Suldan, Z.; Bromleigh, V.; Gamble, M.; Naar, A.M.; Erdjument-Bromage, H.; Tempst, P. and Freedman, L.P. (1999) Ligand-dependent transcription activation by nuclear receptors requires the DRIP complex. *Nature 398*, 824–828.

[42] Pratt, W.B.; Galigniana, M.D.; Harrell, J.M. and DeFranco, D.B. (2004) Role of hsp90 and the hsp90-binding immunophilins in signalling protein movement. *Cellular Signalling 16*, 857-872.

[43] Nishi, M.; Takenaka, N.; Morita, N.; Ito, T.; Ozawa, H. and Kawata, M. (1999) Real-time imaging of glucocorticoid receptor dynamics in living neurons and glial cells in comparison with non-neural cells. *The European Journal of Neuroscience 11*, 1927-1936.

[44] Savory, J.G.A.; Hsu, B.; Laquian, I.R.; Giffin, W.; Reich, T.; Haché, R.J.G. and Lefebvre, Y.A. (1999) Discrimination between NL1- and NL2-mediated nuclear localization of the glucocorticoid receptor. *Molecular and Cellular Biology 19*, 1025-1037.

[45] Black, B.E.; Holaska, J.M.; Rastinejad, F. and Paschal, B.M. (2001) DNA binding domains in diverse nuclear receptors function as nuclear export signals. *Current Biology 11*, 1749-1758.

[46] Holaska, J.M.; Black, B.E.; Love, D.C.; Hanover, J.A.; Leszyk, J. and Paschal, B.M. (2001) Calreticulin is a receptor for nuclear export. *The Journal of Cell Biology 152*, 127-140.

[47] Reichardt, H.M.; Kaestner, K.H.; Wessely, O.; Gass, P.; Schmid, W. and Schutz, G. (1998) Analysis of glucocorticoid signalling by gene targeting. *The Journal of Steroid Biochemistry and Molecular Biology 65*, 111-115.

[48] Kallen J., Schlaeppi J.M., Bitsch F., Filipuzzi I., Schilb A., Riou V., Graham A., Strauss A., Geiser M., and Fournier B. (2004) Evidence for ligand-Independent transcriptional activation of the human estrogen-related receptor alpha: Crystal structure of ERRalpha LBD in complex with PGC-1alpha. *The Journal of Biological Chemistry 279*, 49330-49337.

[49] Ohtake, F.; Takeyama, K.; Matsumoto, T.; Kitagawa, H.; Yamamoto, Y.; Nohara, K.; Tohyama, C.; Krust, A.; Mimura, J.; Chambon, P.; Yanagisawa, J.; Fujii-Kuriyama, Y. and Kato, S. (2003) Modulation of oestrogen receptor signalling by association with the activated dioxin receptor. *Nature 423*, 545-550.

[50] Blaustein, J.D. (2004) Minireview: Neuronal steroid hormone receptors: they're not just for hormones anymore. *Endocrinology 145*, 1075-1081.

[51] Grandics, P.; Miller, A.; Schmidt, T.J. and Litwack, G. (1984) Phosphorylation in vivo of rat hepatic glucocorticoid receptor. *Biochemical and Biophysical Research Communications 120*, 59-65.

[52] Hoeck, W. and Groner, B. (1990) Hormone-dependent phosphorylation of the glucocorticoid receptor occurs mainly in the amino-terminal transactivation domain. *The Journal of Biological Chemistry 265*, 5403-5408.

[53] Krstic, M.D.; Rogatsky, I.; Yamamoto, K.R. and Garabedian, M.J. (1997) Mitogen-activated and cyclin-dependent protein kinases selectively and differentially modulate transcriptional enhancement by the glucocorticoid receptor. *Molecular and Cellular Biology 17*, 3947-3954.

[54] Rogatsky, I.; Logan, S.K. and Garabedian, M.J. (1998) Antagonism of glucocorticoid receptor transcriptional activation by the c-Jun N-terminal kinase. *Proceedings of the National Academy of Sciences of the United States of America 95*, 2050-2055.

[55] Gõttlicher, M.; Heck, S. and Herrlich, P. (1998) Transcriptional cross-talk, the second mode of steroid hormone receptor action. *Journal of Molecular Medicine* 76, 480-489.

[56] Baldwin, A.S. (1996) The NF-κB and IκB proteins: New discoveries and insights. *Annual Review of Immunology 14*, 649-681.

[57] Shaulian, E. and Karin, M. (2002) AP-1 as a regulator of cell life and death. *Nature Cell Biology 4*, E131-135.

[58] Herrlich, P. (2001) Cross-talk between glucocorticoid receptor and AP-1. *Oncogene 20*, 2465-2475.

[59] Auphan, N.; DiDonato, J.; Rosette, C.; Helmbertg, A. and Karin, M. (1995) Molecular basis for immunosuppression by glucocorticoids: inhibition of NF-κB activity through induction of IκB synthesis. *Science 270*, 286-290.

[60] Scheinman, R.; Cogswell, P.; Lofquist, A. and Baldwin, A. (1995) Role of transcriptional activation of IκBα in mediation of immunosuppression by glucocorticoids. *Science 270*, 283-286.

[61] De Bosscher, K.; Berghe, W.V. and Haegeman, G. (2000) Mechanisms of anti-inflammatory action and of immunosuppression by glucocorticoids: negative interference of activated glucocorticoid receptor with transcription factors. *Journal of Neuroimmunology 109*, 16-22.

[62] De Bosscher, K.; Berghe, W.V.; Vermeulen, L.; Plaisance, S.; Boone, E. and Haegeman, G. (2000) Glucocorticoids repress NF-κB-driven genes by disturbing the interaction of p65 with the basal transcription machinery, irrespective of coactivator levels in the cell. *Proceedings of the National Academy of Sciences of the United States of America 97*, 3919-3924.

[63] Assenat, E.; Gerbal-Chaloin, S.; Larrey, D.; Saric, J.; Fabre, J.M.; Maurel, P.; Vilarem, M.J. and Pascussi, J.M. (2004) Interleukin 1beta inhibits CAR-induced expression of hepatic genes involved in drug and bilirubin clearance. *Hepatology 40*, 951-960.

[64] Kamei, Y.; Xu, L.; Heinzel, T.; Torchia, J.; Kurokawa, R.; Gloss, B.; Lin, S.C.; Heyman, R.A.; Rose, D.W.; Glass, C.K. and Rosenfeld, M.G. (1996) A CBP integrator complex mediates transcriptional activation and AP-1 inhibition by nuclear receptors. *Cell 85*, 403-414.

[65] Shimojo, M.; Hiroi, N.; Yakushiji, F.; Ueshiba, H.; Yamaguchi, N. and Miyachi, Y. (1995) Differences in down-regulation of glucocorticoid receptor mRNA by cortisol, prednisolone and dexamethasone in HeLa cells. *Endocrine Journal 42*, 629-636.

[66] Bellingham, D.L.; Sar, M. and Cidlowski, J.A. (1992) Ligand-dependent down-regulation of stably transfected human glucocorticoid receptors is associated with the loss of functional glucocorticoid responsiveness. *Molecular Endocrinology 6*, 2090-2102.

[67] Wallace, A.D. and Cidlowski, J.A. (2001) Proteasome-mediated glucocorticoid receptor degradation restricts transcriptional signaling by glucocorticoids. *The Journal of Biological Chemistry 276*, 42714-42721.

[68] Nebert, D.W.; Roe, A.L.; Dieter, M.Z.; Solis, W.A.; Yang, Y. and Dalton, T.P. (2000) Role of the Aromatic Hydrocarbon Receptor and [Ah] Gene Battery in the Oxidative Stress Response, Cell Cycle Control, and Apoptosis. *Biochemical Pharmacology 59*, 65–85.

[69] Hahn, M.E. (2002) Aryl hydrocarbon receptors: diversity and evolution. *Chemico-Biological Interactions 141*, 131–160.

[70] Mimura, J. and Fujii-Kuriyama, Y. (2003) Functional role of AhR in the expression of toxic effects by TCDD. *Biochimica et Biophysica Acta 1619*, 263–268.

[71] Matsumura, F. (2003) On the significance of the role of cellular stress response reactions in the toxic actions of dioxin. *Biochemical Pharmacology 66*, 527–540.

[72] Safe, S. (2001) Molecular biology of the Ah receptor and its role in carcinogenesis. Toxicology Letters 120, 1–7.

[73] Denison, M.S.; Pandini, A.; Nagy, S.R.; Baldwin, E.P. and Bonati, L. (2002) Ligand binding and activation of the Ah receptor. *Chemico-Biological Interactions 141*, 3 –24.

[74] Denison, M.S. and Nagy, S.R. (2002) Activation of the aryl hydrocarbon receptor by structurally diverse exogenous and endogenous chemicals. *Annual Review of Pharmacology and Toxicology 43*, 309-334.

[75] Barouki, R. and Morel, Y. (2001) Repression of cytochrome P450 1A1 gene expression by oxidative stress: mechanisms and biological implications. *Biochemical Pharmacology 61*, 511-516.

[76] Delescluse, C.; Lemaire, G.; de Sousa, G. and Rahmani, R. (2000) Is CYP1A1 induction always related to AHR signaling pathway? *Toxicology 153*, 73–82.

[77] Hodek, P.; Trefil, P. and Stiborova, M. (2002) Flavonoids-potent and versatile biologically active compounds interacting with cytochromes P450. *Chemico-Biological Interactions 139*, 1-21.

[78] Pollenz, R.S. (2002) The mechanism of AH receptor protein down-regulation (degradation) and its impact on AH receptor-mediated gene regulation. *Chemico-Biological Interactions 141*, 41–61.

[79] Honkakoski, P. and Negishi, M. (2000) Regulation of cytochrome P450 (CYP) genes by nuclear receptors. *The Biochemical Journal 347*, 321-337.

[80] Villard, P.H.; Fallone, F.; Sérée, E.; Rimet, O.; Bourgarel-Rey, V.; Fouchier, F.; Barra, Y.; Durand, A. and Lacarelle, B. (2004) Repressive effect of retinoic acid toward the Ah receptor CYP1A1 induction pathway. *Fundamental and Clinical Pharmacology 18*, p. 219. 8th Annual Meeting of the Societe Francaise de Pharmacologie, Strasbourg, 26-28 April *2004*.

[81] Puga, A.; Xia, Y.and Elferink, C. (2002) Role of the aryl hydrocarbon receptor in cell cycle regulation. *Chemico-Biological Interactions 141*, 117–130.

[82] Vorgias, C.E.; Perides, G.A.; Traub, P. and Sekeris, C.E. (1988) Colchicine, colcemide and cytochalasin B do not affect translocation of the glucocorticoid hormone-receptor in rat thymocytes or Ehrlich ascites cells. *Bioscience Reports 8*, 193-197.

[83] Szapary, D.; Barber, T.; Dwyer, N.K.; Blanchette-Mackie, E.J.and Simons, S.S. (1994) Microtubules are not recquired for glucocorticoid receptor mediated gene induction. *The Journal of Steroid Biochemistry and Molecular Biology 51*, 143-148.

[84] Akner, G.; Mossberg, K.; Sundqvist, K.G.; Gustafsson, J.A and Wikstrom, A.C. (1992) Evidence for reversible, non-microtubule and non-microfilament-dependent nuclear translocation of hsp90 after heat shock in human fibroblasts. *European Journal of Cell Biology 58*, 356-364.

[85] Galigniana, M.; Sruggs, J.L.; Herrington, J.; Welsh, M.J.; Carter-Su, C.; Housley, P.R. & Pratt, W.B. (1998) Heat shock protein 90-dependent (geldanamycin-inhibited) movement of the glucocorticoid receptor through the cytoplasm to the nucleus requires intact cytoskeleton. *Molecular Endocrinology 12*, 1903-1913.

[86] Whitesell, L.and Cook, P. (1996) Stable and specific binding of heat shock protein 90 by geldanamycin disrupts glucocorticoid receptor function in intact cells. *Molecular Endocrinology 10*, 705-712.

[87] Galigniana, M.D.; Housley, P.R.; DeFranco, D.B.and Pratt, W.B. (1999) Inhibition of glucocorticoid receptor nucleocytoplasmic shuttling by okadaic acid requires intact cytoskeleton. *The Journal of Biological Chemistry 274*, 16222-16227.

[88] Galigniana, M.D.; Radanyi, C.; Renoir, J.M.; Housley, P.R.and Pratt, W.B. (2001) Evidence that the peptidylprolyl isomerase domain of the hsp90-binding immunophilin FKBP52 is involved in both dynein interaction and glucocorticoid receptor movement to the nucleus. *The Journal of Biological Chemistry 276*, 14884-14889.

[89] Wang, T.H.; Wang, H.S.; Ichijo, H.; Giannakakou, P.; Foster, J.S.; Fojo, T and Wimalasena, J. (1998) Microtubule-interfering agents activate c-Jun N-terminal kinase/stress-activated protein kinase through both Ras and apoptosis signal-regulating kinase pathways. *The Journal of Biological Chemistry 273*, 4928-4936.

[90] Itoh, M.; Adachi, M.; Yasui, H.; Takenawa, M.; Tanaka, H. & Imai, K. (2002) Nuclear export of glucocorticoid receptor is enhanced by c-Jun N-Terminal kinase-mediated phosphorylation. *Molecular Endocrinology 16*, 2382-2392.

[91] Dvořák, Z.; Modrianský, M.; Ulrichová, J.; Maurel, P.; Vilarem, M.J.and Pascussi, J.M. (2005) Disruption of microtubules leads to glucocorticoid receptor degradation in HeLa cell line. *Cellular Signalling 17*, 187-196.

[92] Kolluri, S.K.; Weiss, C.; Koff, A. & Gottlicher, M. (1999) p27^{Kip1} induction and inhibition of proliferation by the intracellular Ah receptor in developing thymus and hepatoma cells. *Genes and Development 13*, 1742-1753.

[93] Santini, R.P.; Myrand, S.; Elferink, C.and Reiners, J.J.Jr. (2001) Regulation of *Cyp1a1* Induction by Dioxin as a Function of Cell Cycle Phase. *The Journal of Pharmacology and Experimental Therapeutics 299*, 718-728.

[94] Hestermann, E.V.; Stegeman, J.J.and Hahn, M.E. (2002) Relationships among the cell cycle, cell proliferation, and aryl hydrocarbon receptor expression in PLHC-1 cells. *Aquatic Toxicology 58*, 201-213.

[95] Scholler, A.; Hong, N.J.; Bischer, P.and Reiners, J.J.Jr. (1994) Short and long term effects of cytoskeleton-disrupting drugs on cytochrome P450 Cyp1a-1 induction in murine hepatoma 1c1c7 cells: suppression by the microtubule inhibitor nocodazole. *Molecular Pharmacology 45*, 944-954.

[96] Abel, G.A.; Wochnik, G.M.; Rüegg, J.; Rouyer, A.; Holsboer, F.and Rein, T. (2002) Activity of the GR in G2 and mitosis. *Molecular Endocrinology 16*, 1352-1366.

[97] Kino, T.; Souvatzoglou, E.; DeMartino, M.U.; Tsopanomihalu, M.; Wan, Y.and Chrousos, G.P. (2003) Protein 14-3-3sigma interacts with and favors cytoplasmic subcellular localization of the glucocorticoid receptor, acting as a negative regulator of the glucocorticoid signaling pathway. *The Journal of Biological Chemistry 278*, 25651-25656.

[98] Wang, Z.and Garabedian, M.J. (2003) Modulation of glucocorticoid receptor transcriptional activation, phosphorylation, and growth inhibition by p27Kip1. *The Journal of Biological Chemistry 278*, 50897-50901.

[99] Kountouras, J.; Boura, P.and Lygidakis, N.J. (2001) Liver regeneration after hepatectomy. *Hepatogastroenterology 48*, 556-562.

[100] Pascussi, J.M.; Gerbal-Chaloin, S.; Pichard-Garcia, L.; Daujat, M.; Fabre, J.M.; Maurel, P.and Vilarem, M.J. (2000) Interleukin-6 negatively regulates the expression of

pregnane X receptor and constitutively activated receptor in primary human hepatocytes. *Biochemical and Biophysical Research Communications 274*, 707-713.

[101] Zimmers, T.A.; McKillop, I.H.; Pierce, R.H.; Yoo, J.Y.and Koniaris, L.G. (2003) Massive liver growth in mice induced by systemic interleukin-6 administration. *Hepatology 38*, 326-334.

[102] Honkakoski P, and Negishi M (1998) Protein serine/threonine phosphatase inhibitors suppress phenobarbital-induced Cyp2b10 gene transcription in mouse primary hepatocytes. *The Biochemical Journal 330*, 889-895.

[103] Celander, M.; Hahn, M.E.and Stegeman, J.J. (1996) Cytochromes P450 (CYP) in the Poeciliopsis lucida hepatocellular carcinoma cell line (PLHC-1): dose- and time-dependent glucocorticoid potentiation of CYP1A induction without induction of CYP3A. *Archives of Biochemistry and Biophysics 329*, 113-122.

[104] Lai, K.P.; Wong, M.H.and Wong, C.K. (2004) Modulation of AhR-mediated CYP1A1 mRNA and EROD activities by 17beta-estradiol and dexamethasone in TCDD-induced H411E cells. *Toxicological Sciences 78*, 41-49.

[105] Kinyamu, H.K.and Archer, T.K. (2003) Estrogen receptor-dependent proteasomal degradation of the glucocorticoid receptor is coupled to an increase in mdm2 protein expression. *Molecular and Cellular Biology 23*, 5867-5881.

[106] Abbott, B.D.; Perdew, G.H.; Buckalew, A.R.and Birnbaum, L.S. (1994) Interactive regulation of Ah and glucocorticoid receptors in the synergistic induction of cleft palate by 2,3,7,8-tetrachlorodibenzo-p-dioxin and hydrocortisone. *Toxicology and Applied Pharmacology 128*, 138-150.

In: Focus on Cellular Signalling Research
Editor: Dorothy T. Leeds, pp. 55-71

ISBN 1-59454-619-3

Chapter III

Analysis of the Formation of Heterotopic Ossifications in a Cell Model – Effects of Irradiation on Cellular Signalling Regarding the BMP-Receptors

***F. Pohl*[1]*, O. Koelbl*[2]*, M. Flentje*[1]*, and A. Nohe*[3]**

[1]Department of Radiotherapy, University of Würzburg, Wuerzburg Germany
[2]Department of Radiotherapy, University of Regensburg, Regensburg, Germany
[3]Chemical and Biological Engineering, Institute of Molecular Biophysics University of Maine, Orono, Maine, USA

Abstract

These results suggest that radiotherapy blocks cellular responsiveness to BMP-2 induced osteoblastic differentiation in a time and dose dependent manner within a short window of time. The primary mechanism seems to be down-regulation of the BMP-2/receptor complex. Further analyses should differentiate between an influence on BMP-receptor structure, BMP-receptor function (binding capacity) or a combination. This effect is clearly different from the way ionizing irradiation acts in cytotoxic radiotherapy. Taking a possible down-regulation of the BMP-2/receptor complex in account, we proposed that the radiation of C2C12 cells influences the BMP receptor signalling by interfering with the formation or reshuffling of the BMP receptors on the cell surface and will cause changes in the localization of the BMP receptors in specific domains by down regulation of proteins specific for BMP regulation.

Introduction

Heterotopic or ectopic ossification (HO) is defined as the formation of new bone within extraskeletal soft tissues, in which under normal circumstances no ossification occurs. Figure

1 displays an ectopic ossification upwards of the upper part of the femur after endoprothesis of the hip that leads to the complete block of movement. The aetiology of the development of ectopic bone can be discerned in two types: the rare hereditary form and the more frequent acquired form.

The *genetic form of ectopic bone* is normally part of the symptom complex of the autosomal dominant Fibrodysplasia ossificans progressive (FOP) that occurs in childhood. Symptoms are a successive symmetric ossification of muscles starting with the back and shoulder musculature. The affected muscles loose their function and the progression of the disease leads to a complete loss of movement, the endpoint of Fibrodysplasia ossificans progressive. Recently, potentially causative mutations for FOP have been mapped to two sites, adding evidence of the BMP (Bone Morphogenetic Protein) role in HO formation. BMPs play crucial roles in skeletal development as well as during all stages of embryonic development. The first mutation site for FOP lies on the long arm of chromosome 17, in the region of the noggin gene (*NOG*), which is a BMP antagonist. The second genetic location is on the long arm of chromosome 4, in the region of a known BMP-signalling pathway gene. Bone morphogenetic protein 4 is overproduced in patients with FOP.

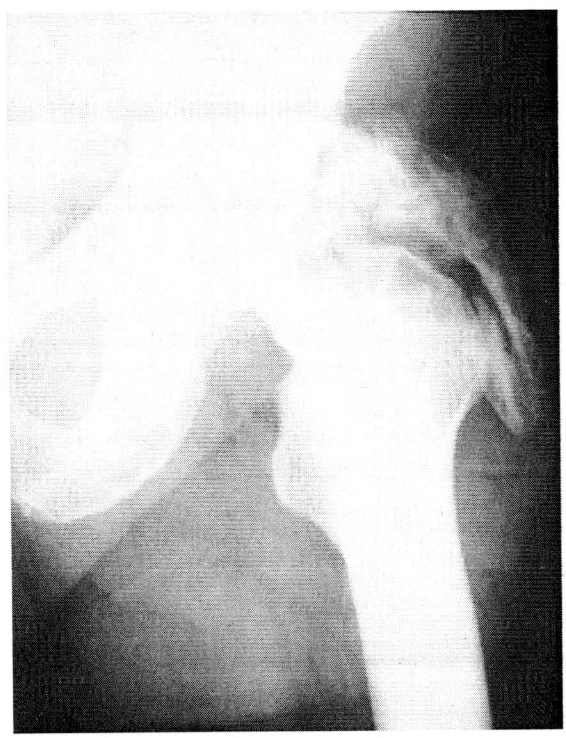

Figure 1. Heterotopic ossification Stage Brooker IV with ankylosis after Total-hip-replacement

The *acquired form of hetereotopic ossification* cuts into three groups named by Bosse [1]:

Non-traumatic ossification. The *non-traumatic form* is seldom found in humans and defined as pseudo-malign bony tumor [2] ot tumor-like lesions. The reason for these group of tumors are supposedly a failure of evolution [3]. The oncologist has to expel the existence of malign tumors like the parosteal osteogenetic sarcoma for example.

Neurologic ossification. The *neurologic form* of heterotopic ossification ossification was firstly described as a "calcification of muscles" in paralysed patients by Goldberg in 1877 [4]. The development of this form can arise in patients with injury to brain and spinal cord, tetanus, poliomyelitis, Guillain-Barré syndrome, and prolonged pharmacologic paralysis during mechanical ventilation, also have been associated with HO formation. The pathogenesis of neurologic ossification is still unknown. Current research points to biotropical factors like immobilisation and the absence of innervation are largely responsible [5]. Moreover, the existence of systemic factors in the tissue induce the differentiation and proliferation of mesenchymal osteogenic precursor cells.

Traumatic ossification. The *traumatic form* of ectopic ossification was discovered by Otto in 1816 [6] and is the most frequent form of ossification today. It is found after extensive fractures and soft tissue injuries. The incidence of heterotopic ossification after fractures near the knee joint is 26% [7] or near the elbow 50% [8]. The development of ectopic bone after burning are rare with an incidence of less than 5 percent and belongs to the traumatic form, too [9]. But the most frequent and clinical relevant ossifications are found after elective or posttraumatic (after acetabulum fracture) endoprosthetic total hip replacement (THR). As THR is the most frequent and clinical relevant form of heterotopic ossification, there is also the best and sucessful research how to prevent THR. But there is still no knowledge about the associated mechanisms underlying this prevention.

In literature the cumulative percentage of heterotopic ossification after total hip replacement is described between 8 and 91% in literature (Table 1). This large range is manifested by means of different risk profiles of patients.

Table 1.Percentage of HO after THR w/o prophylactic therapy

Literature	Year	Number of patients	Percentage of HO in %
Brooker [31]	1973	100	21
Lazansky [32]	1973	501	8
Rosendahl [33]	1977	70	90
Koelbl [34]	1997	100	65

Aetiology

The generally accepted mechanism of the development of heterotopic ossification indicated that primordial mesenchymal cells may be induced by special proteins to differentiate into osteoprogenitor cells. These osteoprogenitor cells modulate then into osteoblasts. The osteoblasts lay down osteoid matrix, which then calcifies to form osteocytes and heterotopic bone [10].

Craven and coauthors could demonstrate in animal studies with tritiated hydrogen labelled thymidine and uridine the transformation of fibroblasts in the soft tissues surrounding an implanted demineralised bone fragment into migrating mesenchymal cells that developed to differentiate into osteoblasts [11].

These observed differentiation into osteoblasts is stimulated by members of the BMP family .a subgroup of the TGF-ß family [12]. These proteins are potent inducers of Bone formnation as it was shown in vivo and in vitro.

Looking at the traumatic form of heterotopic ossification the differentiation from mesenchymal to osteoblastic tissue has been noted to occur by 16 hours after operation, with a peak response noted at approximately 32 hours [35].

Prophylaxis of Heterotopic Ossification

It is very important to keep the described clinical data about the time depending genesis of heterotopic ossification after traumatic events like THR in mind to find a prophylactic treatment for preventing the development of heterotopic ossification, because it seems to be necessary to block the differentiation at a very early stage. Once a formation of ectopic bone has take place only a re-surgery can help.

Clinical studies showed that pre- or postoperative irradiation of the hip region avoids the formation of HO after THR (Table 2).

Table 2. Pre- and postoperative radiotherapy studies for prevention of HO

Author	Year	Hips (n)	Radiation regime	HO in %
Seegenschmiedt [36]	1997	111	postop. 5 x 3,5 Gy 17,5 Gy	6%
Kölbl [37]	1997	101	postop. 4 x 3 Gy→12 Gy	5%
Lonardi [38]	2001	118	preop. 1 x 17,5 Gy	4%
Seegenschmiedt [39]	2001	1116	preop. 1 x 5-7 Gy	8,7%

The molecular mechanisms responsible for the effects of ionizing rays on HO are poorly understood. So far there is no side effect caused by using pre- or postoperative irradiation of the hip region known looking at nearly twenty years of practice. Moreover, nonsteroidal anti-inflammatory drugs (NSAID), taken for up to a minimum of two to three weeks could be used for prophylaxis of heterotopic ossification after THR, but is accompanied by the high risk of gastritis following active gastric ulcer especially in elderly, who are the typical patients needing THR. The possibility of pharmacological prophylaxis was discovered in a study by Dahl [40] and coauthors in 1975.

Comparable to the fragmentary knowledge for the molecular mechanisms responsible for the effects of ionizing rays on HO it is still unclear why NSAID`s inhibit heterotopic ossification. One possible mechanism of NSAID`s may be caused through their ability to inhibit prostaglandin synthesis, which may interfere with the inflammatory response after trauma. Schmidt and coauthors described a connection between this inflammatory response and development of ectopic bone formation [41].

Treatment with ionising radiation is an essential element in the therapy of benign and malignant diseases. However, there is a fundamental difference in used doses and fractionation. The dose and fractionation of radiotherapy for malignant cells aims at cell-death of the malignant cell. Dose and fractionation in the therapy of benign diseases are empirical. Since the first description of radiation treatment for prevention of HO after total hip replacement [45] there are various clinical and animal studies, which analyzed different concepts of dose and fractionation [36, 47, 48].

In these studies not only the radiation dose but also the time interval to the point of surgery was important. Clinical trials using radiation doses lower than 7 Gy showed increased HO-rates, a dose escalation beyond 7 Gy did not result in lower HO-rates [49]. Therefore a threshold for an effective radiation dose of about 7 Gy can be assumed [48, 51]. Furthermore, independent of dose HO was only prevented by radiotherapy in an interval from one day pre- to four days postoperatively. This observation suggests, that the development of ectopic bone can be influenced by radiotherapy only at its beginning, and can not be explained by a cytotoxic effect of radiotherapy.

The cells responsible for the HO are ubiquitous existing mesenchymal pluripotent stem cells [46, 50]. The stimulus inducing the errant osteogenic differentiation of pluripotent cells are BMPs released by the postoperative inflammation. The BMPs, members of the transforming growth factor-ß (TGF-ß) superfamily, were first identified as factors that induce ectopic bone formation in vivo, when implanted into muscular tissues [52, 53]. Using the chicken „limb bud“, an embryonal, mesenchymal tissue equivalent, the influence of BMP on the osteogenic differentiation of mesenchymal cells was shown in vitro [54]. In a recent study we analyzed the influence of radiotherapy on BMP induced differentiation using the chicken limb bud assay [55]. The dose and time dependent effectiveness of radiotherapy known from the results of clinical studies could be reproduced and a cytotoxic effect of radiotherapy was excluded [55]. Due to that observation an interaction between radiotherapy and the BMP induced signal cascade resulting in an osteogenic differentiation of pluripotent cells was assumed.

Bone Morphogenetic Proteins and their Receptors

The formation of the human skeleton during embryonic development is very complex. Individual growth factors generate signals that in turn create gradients in the distribution of different growth factors. In the end, this leads to the selective activation of specific target genes and the development of bone by desmal or chondral ossification. In the adult, bones are still remodeled. Osteoclasts resorb bone, which is restored again by osteoblasts. If the equilibrium between the resorption of bone and the production of new bone is disturbed, diseases such as osteoporosis result. From 1990 to 1992 six growth factors were isolated by extraction of bovine bone [13, 14, 15, 16].

These growth factors, called BMP's, induce bone formation when injected into the muscle of mice. Some members of the BMP subgroup of proteins called Growth Differentiation Factors (GDF), including GDF-5, GDF-6 and GDF-7, are BMP like proteins but others have been only partially characterized. The BMP's belong to the Cystin-knot family and form a subgroup of the Transforming Growth Factor (TGF)-beta superfamily. There are over 30 proteins in the BMP subgroup. They play important roles not only in the development of bone, but also during gastrulation, neurogenesis, chondrogenesis and the development of the inner organs and are potential bone formation inducing proteins [13, 17, 18, 19, 20, 21].

Three BMP-receptors and three activin receptors have been shown to mediate signalling between the extracellular and cytoplasmic compartments, but for some BMP's the receptor interaction is not known yet. The receptors are Serine/Threonine kinases and are classified as type-I (BRI) or type-II (BRII) receptors. The type-II receptor is a constitutively active kinase. The type-I receptors have a GS-box domain and are phosphorylated and activated by the type-II receptors. The signalling cascade is activated only when the BMP's bind to both the type-I and type-II receptors [22, 23, 24, 35, 26, 27, 28]. Therefore a crucial step for signal activation is the binding of BMP to at least one type-I and one type-II receptor [29,30].

Upon binding of BMP to its receptors phosphorylation of Smad1, -5 and -8 (R-Smads) is triggered. The activated phospho-R-Smads bind to SMAD4 and this complex moves to the nucleus, causing the activation of BMP-targetgenes. Recent data show that the induction of ALP by BMP-2 is triggered by a Smad-independent pathway (initiated through a oligomerization of the receptors induced by BMP-2. BMP-2 binds to BRI with high affinity, which recruits BRII) via MAPKinase (*m*itogen *a*ctivated *p*rotein) p38 (Figure 2) [42].

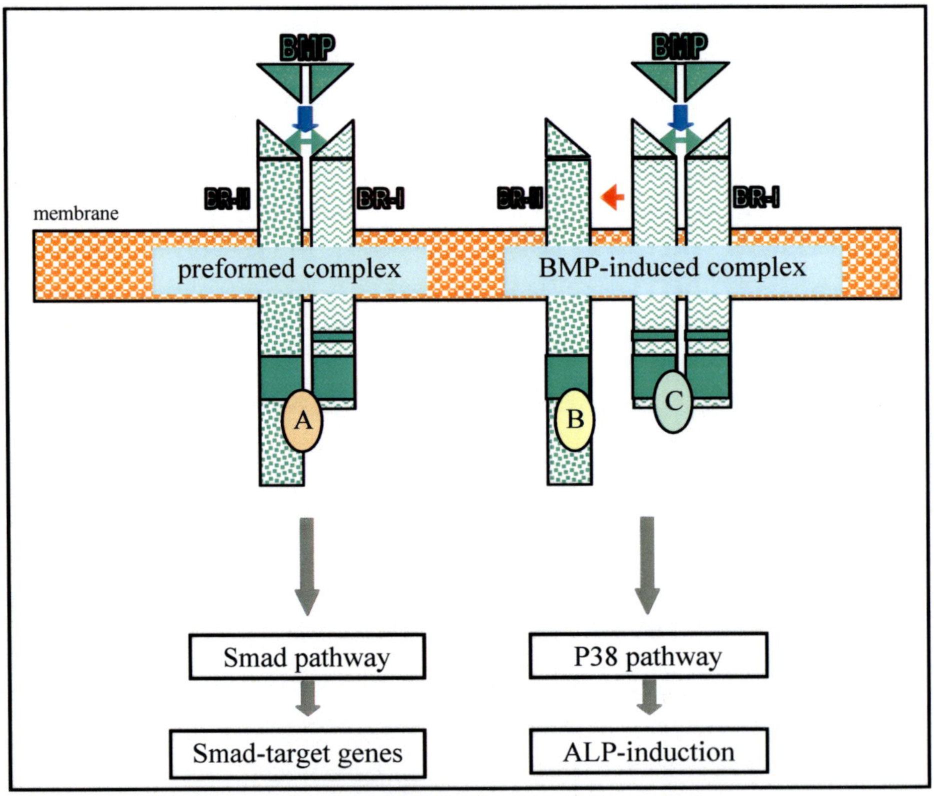

Figure. 2. Receptor-activation and cell signalling through BMP-2 receptors

Cell Model

To analyse the molecular mechanisms leading to the development of ectopic bone we designed a suitable cell model with C2C12 cells out of mice. C2C12 cells are pluripotent mesenchymal fibroblastic precursor cells. In vitro the cells show post-confluence a formation of myotubular structures and secretion of muscle-specific proteins. Under physiological conditions there is no measurable alkaline phosphatase activity. However incubation with Bone Morphogenetic Protein (BMP-2) leads to formation of nodules (as in endochondral osteogenesis) and to secretion of Alkaline Phosphatase (ALP) and osteocalcin production (i.e. differentiation of the cell in direction osteoblast, bone formation). ALP-activity in these cells acts as marker for differentiation and proliferation of these myoblastic progenitor cells towards osteoblasts as described by Tylzanowski et al. [16]. Irradiation pre-induction with BMP-2 causes an decrease of the alkaline phosphatase activity. The pluripotent mesenchymal precursor cell line C2C12 provides a model system to study the osteoblast differentiation during bone formation in muscular tissues and to analyze the point of interaction between radiotherapy and the cellular/intracellular signal pathway [56].

Influence of BMP Receptor Distribution on Signalling

Recent work by Nohe and coworkers [30] show that the BMP-receptors can exist as preformed homo- and hetero-oligomers on the cell surface. They transfected plasmids encoding the BMP-receptors into COS and C3H10T1/2 cells and showed that BRII formed ligand independent complexes with BRIa and BRIb and that BRIa also formed complexes with BRIb. Furthermore, BRII, BRIa and BRIb each formed homo-oligomeric complexes. Addition of the ligand (BMP-2) caused an increase in the number of hetero-oligomeric complexes (BRII with BRIb or BRIa) and homo-oligomeric complexes (BRIa or BRIb). The number of homo-oligomeric complexes of BRII did not increase because of the low binding affinity of BMP-2 to BRII in the absence of BR-I. These results show a very flexible oligomerization pattern of the BMP-receptors at the cell surface even in the absence of BMP's. The changes in the distribution of type-I and type-II receptors between homo- and hetero-oligomeric complexes may provide a mechanism for specificity of BMP signalling. To determine the function of the preformed hetero-oligomeric complexes of BRII and BRIb or BRIa receptors at the cell surface, truncations of BRII were transfected into COS cells either alone or with BRIa or BRIb. Immunoprecipitation studies followed by Western-Blotting and also co-patching studies revealed that one mutant, TC3, which lacks the C-terminal part of BRII downstream of the kinase domain, can form a preformed hetero-oligomeric complex with BRIb but not with BRIa. Reporter gene assays in C2C12 or MC3T3 cells using a reporter gene construct that is sensitive to activation of the Smad pathway, demonstrated the ability of BRIa and TC3 but not BRIb and TC3 to activate the Smad signalling cascade. BRIa and TC3 were still able to activate the p38 pathway as measured by p38 phosphorylation. The results obtained with these BRII mutants give evidence that there are two different signalling pathways: one requiring the existence of preformed complexes and one using the homo-oligomeric complexes [29, 30, 43]. These results show once again the fine regulation of the BMP's by distinct receptor complexes.

Further data indicate that the co-expression of the BMP-type-II receptor (BRII) influences the aggregation and the distribution of the BMP-type-Ia receptor (BRIa) in COS7 cells and in A431 cells. BMP-2 stimulation of the cells leads to a rearrangement of receptor complexes at the cell surface. Using A431 cells and limb bud-derived mesenchymal cells it is shown that co-expression of the BRII and a constitutive active BRIa-ca is necessary for the activation of the Smad pathway. More importantly, when using a kinase deficient BMP-type-II receptor the rearrangement of BRIa is blocked. Together, these findings suggest that the rearrangement of the receptors at the cell surface prior to forming preformed ligand independent complexes plays a critical role in activation of the Smad pathway. It also suggests further that the kinase activity of BRII is needed for signaling beyond the activation of BRIa at the GS domain. This study also indicates further that the kinase activity of BRII plays a crucial role in the activation and reshuffling process. These data once again indicate the fine tuning and importance of BMP aggregation and clustering on the cell surface level.

Distribution and Co-Localization of Receptors at the Cell Surface

In addition to the described standard biochemical and cell biological experimental approaches, we used Image Correlation Spectroscopy and Image Cross-Correlation Spectroscopy to determine the distribution and co-localization of receptors at the cell surface.

Image correlation spectroscopy (ICS) provides quantitative information about the average number and size of clusters of a single protein at the cell surface per unit area. Image cross-correlation spectroscopy (ICCS) provides quantitative information about the fraction of one protein that is co-localized with a second protein [62]. Both ICS and ICCS rely on calculating correlation functions from confocal images collected at the highest possible resolution and have been applied successfully to a number of membrane receptor systems. For example, Petersen and coworkers have demonstrated that ICS and ICCS are very powerful techniques for studying the association of receptors with specific domains at the cell surface and that the percentage of receptors present in these domains, e.g. caveolae or coated pits, can be quantitatively measured [62-70].

ICS and ICCS: In addition to standard biochemical and cell biological experimental approaches, we intend to use Image Correlation Spectroscopy (ICS) and Image Cross-Correlation Spectroscopy to determine the distribution and co-localization of receptors at the cell surface. These techniques were invented in our laboratory [62-70] and have been applied to studies of numerous receptor systems. ICS relies on collecting high magnification images (15x15 μm) with a confocal microscope. The intensity distribution will reflect the distribution of the protein of interest. A uniform intensity across the image reflects a uniform distribution of protein – a large number of clusters or monomers. A heterogeneous intensity, in turn, reflects proteins in highly aggregated states – in clusters. We have shown that the amplitude of the autocorrelation function of the image is inversely proportional to the average number of clusters per unit area. In fact, we can calculate the average cluster density (CD) from the amplitude. The experiments require correction for non-specific binding and background detector noise, but this is straightforward if the proper controls for primary and secondary antibody labeling are included. The CD data are normalized for ease of representation, but the absolute values are available and meaningful. In order to quantify molecular co-localization, we simultaneously label two proteins of interest with distinguishable fluorescent tags and measure the two images of the two chromophores. The autocorrelation functions of each image provide the CD values for each protein separately. The amplitude of the cross-correlation function [70] is known to be proportional to the average number of clusters with both chromophores present, which can yield a density of those clusters that contain both proteins. Calculation of the appropriate ratios, will give an estimate of the fraction of clusters of one protein that contain the other protein. A fraction of 1 corresponds to 100% co-localization, while a fraction of 0.3 would reflect 30% of those clusters having the other protein in them. We, therefore, have the tools to quantitatively estimate the number of receptor clusters and the extent of co-localization of two different receptors. The experimental approaches, the analysis, the limitations and the pitfalls in measurements and interpretation have been documented extensively and reviewed. The ICS and ICCS experiments are the critical tools for determining receptor distributions and interactions and how these change with treatment or mutations. Nohe and co-workers used this approach to investigate the distribution of the receptors in certain membrane domains. The data obtained indicate that the receptors need to shuttle on the cell surface between distinct domains in order to signal. This shows a very flexible and dynamic pattern of the receptors, providing evidence for the regulation of BMP signalling within the plasma membrane.

Influence of Radiation on BMP Signalling

Recent data on an analogue model by Pohl, Koelbl and coworkers [44] show that the BMP-receptors can exist as preformed homo- and hetero-oligomers on the cell surface. They show a time and dose dependent effect of irradiation. In C2C12 cells the osteogenic differentiation and the BMP signalling pathways are influenced by ionizing radiation. With increased dose of irradiation from 0 to 20 Gy the signal of bound and crosslinked BMPR-Ia decreased (Figure 3). Whether this process results from down-regulation of the amount of the BMP-receptors and/or by changes in the binding capability of the ligand to its receptor can not be answered by this experiment.

Activated BMPR-I receptors phosphorylate Smad1, Smad5, and Smad8, which then assemble into heteromeric complexes with Smad4 [58]. It can be assumed that the BMP-2/receptor complex reduced by radiotherapy should result in a decreased Smad phosphorylation. To examine this hypothesis, we analyzed whether radiotherapy inhibits phosphorylation of Smad1. Treatment of cells with BMP-2 increased Smad1 phosphorylation (Figure 4).

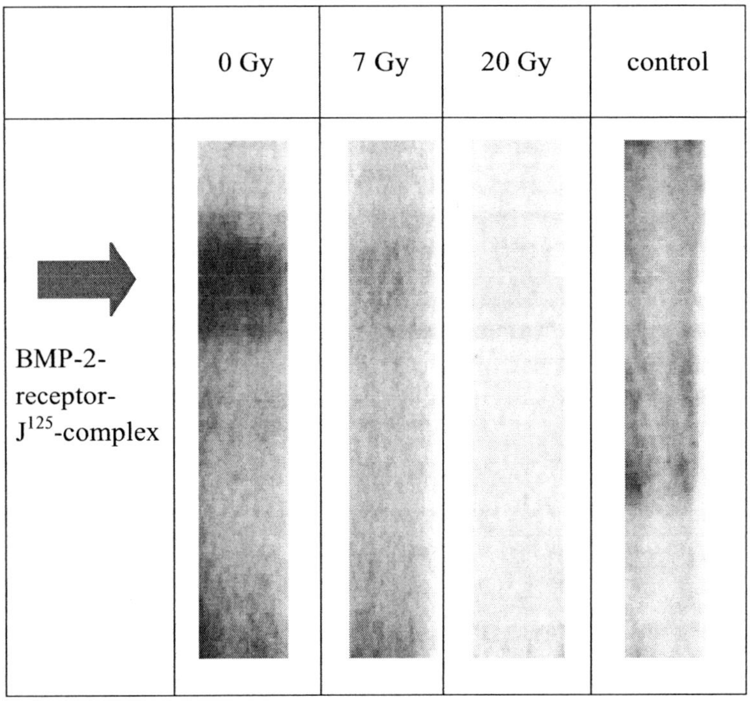

Figure 3. Detection of cell surface BMP-2-receptors by binding and chemical crosslinking of ^{125}J-BMP-2

However, pretreatment of cells with radiotherapy before addition of BMP-2 resulted in a time-dependent inhibition of Smad1 phosphorylation. Although BMP-2 induced Smad1 phosphorylation was not sufficiently changed with 2 to 24 hours after incubation, it was almost completely blocked at 38 and 44 hours. In contrast, total Smad1 levels were not affected by radiotherapy (Figure 4). This delayed effective of ionizing irradiation in reducing phosphorylation has its analyze in UV irradiation as described by Quan [57].

Phosphorylated Smad1 normally transfers into the nucleus to regulate transcription of target genes and to initiate the differentiation into osteoblastic cells [46]. Alkaline phosphatase (ALP) activity is a characteristic of osteoblast phenotype and can be induced by BMP-2 in C2C12 cells. Although ALP is an ubiquitous enzyme, ALP activity is widely used as an index of osteoblast differentiation because the activity characteristically increases during early development in osteoblasts [59]. Ikeda et al. analyzed the effect of radiotherapy on BMP induced ALP activity in C2C12 cells [60]. After incubation with BMP-2 an irradiation with 4 or 8 Gy on culture day 2 did not decrease the ALP activity in irradiated cells compared to the unirradiated control. Ikeda et al. concluded that BMP receptors and downstream signal transduction are not influenced by radiotherapy, but he did not consider the time dependent effect of radiotherapy. In our study we confirmed that radiotherapy 24 hours after BMP-2 incubation has no effect on ALP-activity. Radiotherapy within 6 hours before or after incubation however decreased ALP activity significantly (Figure 5).

	1	2	3	4	5	6	7	8	9	10	11	12
	0 h + BMP	0 h - BMP	2 h + BMP	2 h - BMP	14 h + BMP	14 h - BMP	24 h + BMP	24 h - BMP	38 h + BMP	38 h - BMP	44 h + BMP	44 h - BMP
Smad												
Phospho -Smad												

Figure 4. Analyzing the signalling pathway of BMP-2

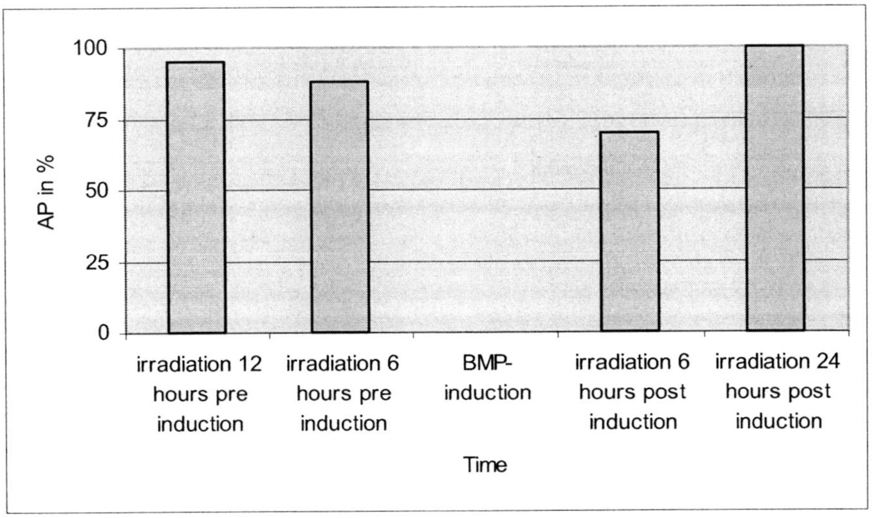

Figure 5. Measurement of ALP (mean) in irradiated and non-irradiated C2C12 cells

In a recent study additionally to the known BMP receptor – Smad pathway a second Smad independent pathway was described [42]. It is initiated through a BMP-2 induced oligomerization of the receptors. The activation of this complex results especially in the induction of ALP. The existence of two different pathways could explain the observation that a reduction of ALP activity was measured even before measuring a reduction of Smad1.

Some authors suggest that the effect of irradiation in prevention of HO is caused primarily by cell damage [61]. Our results measuring cell proliferation and viability as well as Ikeda et al. [60] measuring apoptosis in C2C12 cells after 8 Gy did not found any evidence of cell death within the relevant time window.

In conclusion our results suggest that radiotherapy blocks cellular responsiveness to BMP-2 induced osteoblastic differentiation in a time and dose dependent manner within a short window of time. The primary mechanism seems to be down-regulation of the BMP-2/receptor complex. Further analyses should differentiate between an influence on BMP-receptor structure, BMP-receptor function (binding capacity) or a combination. This effect is clearly different from the way ionizing irradiation acts in cytotoxic radiotherapy.

Taking a possible down-regulation of the BMP-2/receptor complex in account, we proposed that the radiation of C2C12 cells influences the BMP receptor signalling by interfering with the formation or reshuffling of the BMP receptors on the cell surface and will cause changes in the localization of the BMP receptors in specific domains by down regulation of proteins specific for BMP regulation.

Prospect

We could show that the primary mechanism of radiotherapy is an influence on cellular responsiveness to BMP. The chosen cell model allows the analysing of effects of irradiation on the level of cell signalling.

References

[1] Bosse A. Clinical aspects, differential diagnosis and histogenesis of heterotopic ossification. *Veroff Pathol.* 1997;146:1-168

[2] Fine G, Stout AP. Osteogenic sarcoma of the extra-skeletal soft tissues. *Cancer* 9, 1027,1956.

[3] Schaser KD, Bail HJ, Haas NP, Melcher I. Treatment concepts of benign bone tumors and tumor-like bone lesions. *Chirurg.* 2002 Dec; 73(12):1181-90.

[4] Eulert J, Knelles D, Barthel Th. *Orthopäde* 26. 399-406,1997.

[5] Garland DE. A clinical perspective on common forms of acquired heterotopic ossification. Clin Orthop 263, 13-29, 1991

[6] Binnie. On myositis ossificans traumatica. *Ann Surg* 38, 423-440, 1903.

[7] Stannard JP, Wilson TC, Sheils TM, *et al.* Heterotopic ossification associated with knee dislocation. *Arthroscopy* 2002;18:835-839

[8] Ilahi OA, Strausser DW, Gabel GT. Post-traumatic heterotopic ossification about the elbow. *Orthopedics* 1998;21: 265-268.

[9] Richards AM, Klaassen MF. Heterotopic ossification after severe burns: a report of three cases and review of the literature. *Burns* 1997;23:64-68

[10] Plasmans CM, Kuypers W, Slooff TJ. The effect of ethane-1-hydroxy-1, 1-diphosphonic acid EHDP on matrix induced ectopic bone formation. *Clin Orthop.* 1978 May; 132:233-43.

[11] Craven PL, Urist MR. Osteogenesis by radioisotope labelled cell populations in implants of bone matrix under the influence of ionizing radiation. *Clin Orthop.* 1971 May;76:231-3

[12] Vinall RL, Reddi AH. The effect of BMP on the expression of cytoskeletal proteins and its potential relevance. *J Bone Joint Surg* Am. 2001;83-A Suppl 1Pt 1:S63-9

[13] Celeste, A. J., Iannazzi, J. A., Taylor, R. C., Hewick, R. M., Rosen, V., Wang, E. A., Wozney, J. M. (1990). Identification of transforming growth factors present in bone. *Proc. Natl. Acad. Sci. U. S. A.* 87: 9843-9847.

[14] Wang, E. A., Rosen, V., Cordes, P., Hewick, R. M., Kritz, M. J., Luxenberg, D. P., Sibley, B. S., Wozney, J. M. (1988). Purification and characterization of other distinct bone inducing factors. *Proc. Natl. Acad. Sci. U. S. A.* 85: 9484-9488.

[15] Wozney, J. M., Rosen, V., Celeste, A. J., Mitsock, L. M., Whitters, M. J., Kriz, R. W., Hewick, R. M., Wang, E. A. (1988). Novel regulators of bone formation: molecular clones and activities. *Science* 242: 1528-1534.

[16] Wozney, J. M. (1989). Bone morphogenetic proteins. *Prog Growth Factor Res.*1: 267-80.

[17] Cunningham, N. S., Jenkins, N. A., Gilbert, D. J., Copeland, N. G., Reddi, A. H., Lee, S. J. (1995). Growth differentiation factor 10: Anew member of transforming growth factor-ß superfamily related to bone morphogenetic protein-3. *Growth Factors* 12: 99-109

[18] Dunn, N. R., Winnier, G. E., Hargett, L. K., Schrick, J. J., Fogo, A. B., Hogan, B. L. (1997). Haploinsufficient phenotypes in Bmp4 heterozygous null mice and modification by mutations in Gli3 and Alx4. *Dev. Biol.* 188: 235-247.

[19] Padgett, R. W., St. Johnson, R. D., Gelbert, W.M. (1987). A transcript from a Drosophila pattern gene predict a protein homologous to the transforming growth factor-ß family. *Nature* 325: 81-84.

[20] Reddi, H. (1995). Bone morphogenetic proteins. *Adv. Dent. Res.*, 13: 657-683.

[21] Storm, E. E., Huynh, T. V., Copeland, N. G., Jenkins, N. A., Kingsley, D. M., Lee, S. J. (1994). Limb alterations in brachypodism mice due to mutations in a new member of the TGF beta-superfamily. *Nature* 368: 639-643.

[22] Hoodless, P.A., Haerry, T., Abdollah, S., Stapleton, M., O'Connor, M. B., Attisano, L. and Wrana, J. L. (1996). MADR1, a MAD-related protein that functions in BMP2 signalling pathways. *Cell* 85: 489-500.

[23] Liu, F., Ventura, F., Doody, J., Massague, J. (1995). Human type II receptor for bone morphogenic proteins (BMP's): extension of the two-kinase receptor model to the BMP's. *Mol. Cell. Biol.* 15: 3479-3486.

[24] Luo, G., Hofman, C., Bronkers, A. L. J. J., Sohoki, M., Bradley, A., Karsenty, G. (1995). BMP-7 is an inducer of nephrogenesis and is also required for eye development and skeletal patterning. *Genes Dev.* 9: 2808-2820.

[25] Kawabata, M., Chytil, A., Moses, H.L. (1995), Cloning of a novel type II serine/threonine kinase receptor through interaction with the type I transforming growth factor-beta receptor. *J Biol Chem.* 270: 5625-5630.

[26] Macias-Silva, M, Hoodless, P. A., Tang, S. J., Buchwald, M., Wrana, J. L. (1998) Specific activation of Smad1 signalling pathways by the BMP7 type I receptor, ALK2. *J. Biol. Chem.* 273:25628-36.

[27] Nohno, T., Ishikawa, T., Saito, T., Hosokawa, K., Noji, S., Wolsing, D. H., Rosenbaum, J. S. (1995). Identification of a human type II receptor for bone morphogenetic protein-4 that forms differential heteromeric complexes with bone morphogenetic protein type I receptors. *J. Biol. Chem.* 270: 22522-22526.

[28] Rosenzweig, B. L., Imamura, T., Okadome, T., Cox, G. N., Yamashita, H., ten Dijke, P., Heldin, C. H., Miyazono, K. (1995). Cloning and characterization of a human type II receptor for bone morphogenetic proteins. *Proc. Natl. Acad. Sci. U. S. A.* 92: 7632-7636.

[29] Gilboa, L., Nohe, A., Geissendorfer, T., Sebald, W., Henis, Y. I., Knaus, P., (2000). Bone morphogenetic protein receptor complexes on the surface of live cells: a new oligomerization mode for serine/threonine kinase receptors. *Mol. Biol. Cell.* 3:1023-35.

[30] Nohe A., Hassel S., Neubauer, F., Sebald, W., and Knaus. P., (2000). Ligandenunabhaengige Heterooligomere der BMP Rezeptoren sind Vorraussetzung für die Aktivierung des Smad-Pathways. *Osteologie* 9: 66.

[31] Brooker AF, Bowerman JW, Robinson RA, et al: Ectopic ossification following total hip replacement: Incidence and a method of classification. *J Bone Joint Surg* 1973; 55 A: 1629 –1632

[32] Lazansky MG. Complications revisited. The debit side of total hip replacement. Clin Orthop. 1973 Sep;95:96-103

[33] Rosendahl S, Christoffersen JK, Norgaard M. Para-articular ossification following hip replacement. 70 arthroplasties ad modum Moore using McFarland's approach. *Acta Orthop Scand.* 1977;48 (4):400-4.

[34] Koelbl O, Knelles D, Barthel T, *et al.* Preoperative irradiation versus the use of nonsteroidal anti-inflammatory drugs for prevention of heterotopic ossification following total hip replacement: the results of a randomized trial. *Int J Radiat Oncol Biol Phys* 1998; 42: 397-401

[35] Tonna BJ, Cronkite EP. Auto radiographic studies of cell proliferation in the peiosteum of intact and fractured femora of mize utilizing DNA labelling with H3 thymidine. *Proc Soc Exp Biol Med* 107:719, 1961

[36] Seegenschmiedt MH, Keilholz L. Martus P, et al: Prevention of heterotopic ossification about the hip: Final results of two randomized trials in 410 patients using either preoperative or postoperative radiation therapy. *Int J Radiat Oncol Biol Phys* 1997; 39: 161 – 171

[37] Kölbl O, Knelles D, Barthel T, Kraus U, Flentje M, Eulert J. Randomized trial comparing early postoperative irradiation vs. the use of nonsteroidal antiinflammatory drugs for prevention of heterotopic ossification following prosthetic total hip replacement. *Int J Radiat Oncol Biol Phys.* 1997 Dec 1;395:961-6

[38] Lonardi F, Gioga G, Coeli M, Ruffo P, Agus G, Pizzoli A, Campostrini F. Preoperative, single-fraction irradiation for prophylaxis of heterotopic ossification after total hip arthroplasty. *Int Orthop.* 2001;25 (6):371-4.

[39] Seegenschmiedt MH, Makoski HB, Micke O; German Cooperative Group on Radiotherapy for Benign Diseases Radiation prophylaxis for heterotopic ossification about the hip joint--a multicenter study. *Int J Radiat Oncol Biol Phys.* 2001 Nov 1;51(3):756-65

[40] Dahl, H.K.: Klinske Observasjoner. In: *Symposium on Arthose*, Ed Msd Blindern Norway 1975, 37-46

[41] Schmidt SA, Kjaersgaard-Andersen P, Pedersen NW, Kristensen SS, Pedersen P, Nielsen JB. The use of indomethacin to prevent the formation of heterotopic bone after total hip replacement. A randomized, double-blind clinical trial *J Bone Joint Surg Am.* 1988 Jul; 706:834-8.

[42] Nohe A, Hassel S, Ehrlich M, Neubauer F., Sebald W., Henis Y.I., P. Knaus P, The Mode of Bone Morphogenetic Protein (BMP) Receptor Oligomerization Determines Different BMP-2 Signaling Pathways. *J Biol Chem* 277, 5330-5338 (2002)

[43] Nohe A., Keating E., Underhill T. M., Knaus P. and Petersen N. O. (2003). Effect of the distribution and clustering of the type I A BMP receptor (ALK3) with the type II BMP receptor on the activation of signalling pathways. *JCS* 116, 3277-3284

[44] Pohl F, Hassel S, Nohe A, Flentje M, Knaus P, Sebald W, Koelbl O (2003). Radiation-induced suppression of the Bmp2 signal transduction pathway in the pluripotent mesenchymal cell line C2C12: an in vitro model for prevention of heterotopic ossification by radiotherapy. *Radiat Res.* 2003 Mar;159 (3):345-50.

[45] MB, Scanlon PW: The use of radiation to discourage ectopic bone: A nine-year study in surgery about the hip. *J Bone Joint Surg* 1981; 63: 201 – 208

[46] Owen M.. The origin of bone cells in the postnatal organism. *Arthr Rheum* 23, 1073-1080 (1980)

[47] Nishimura R., Kato Y., Chen D., Harris S.E., Mundy G.R., Yoneda T., Smad5 and DPC4 are key molecules in mediating BMP-2-induced osteoblastic differentiation of

the pluripotent mesenchymal precursor cell line C2C12. *J Biol Chem* 273, 1872-1879 (1998)

[48] Seegenschmiedt M.H., Makoski H.B., Micke O., Radiation prophylaxis for heterotopic ossification about the hip joint--a multicenter study. *Int J Radiat Oncol Biol Phys* 51, 756-765 (2001).

[49] Knelles D, Barthel T, Karrer A, *et al.* Prevention of heterotopic ossification after total hip replacement. A prospective, randomised study using acetylsalicylic acid, indomethacin and fractional or single-dose irradiation. *J Bone Joint Surg Br* 1997; 79: 596-602

[50] Friedenstein M.H., Kuralesova A.I. Osteogenic precursor cells of bone marrow in radiation chimeras. *Transplantation* 12, 99-108 (1971)

[51] Hedley A.K., Mead L.P., Hendren D.H., The prevetion of heterotopic bone formation following total hip arthroplasty using 600 rad in a single dose. *J Arthroplasty* 4, 319-325 (1989)

[52] Wozney J.M., Rosen V., Celeste A.J., Mitsock L.M., Whitters M.J., Kriz R.W., Hewick R.M., Wang E.A., Novel regulators of bone formation: molecular clones and activities. *Science* 242, 1528-1534 (1988)

[53] Wang E.A., Rosen V., D'Alessandro J.S., Bauduy M., Cordes P., T. Harada, D.I. Israel, Hewick R.M., Kerns K.M., Wozney J.M., Recombinant human bone morphogenetic protein induces bone formation. *Proc Natl Acad Sci* U S A 87, 2220-2224 (1990)

[54] Carrington J.L., Reddi A.H. Temporal changes in the response of chick limb bud mesodermal cells to transforming growth beta-type 1. *Exp Cell Res* 186 368-373 (1990)

[55] Koelbl O., Knaus P., Pohl F., Flentje M., Sebald W, Irradiation-induced reduction of BMP-induced proteoglycan synthesis in an embryonal mesenchymal tissue equivalent using the chicken "limb bud" test. *Strahlenther Onkol* 177, 432-436 (2001)

[56] Katagiri T., Yamaguchi A., Komaki M., Abe E., Takahashi N., Ikeda T., Rosen V., Wozney J.M., Fujisawa-Sehara A., Suda T., Bone morphogenetic protein-2 converts the differentiation pathway of C2C12 myoblasts into the osteoblast lineage. *J Cell Biol* 127, 1755-1766 (1994)

[57] Quan T., He T., Voorhees J.J., Fisher G.J. Ultraviolet irradiation blocks cellular responses to transforming growth factor-beta by down-regulating its type-II receptor and inducing Smad7. *J Biol Chem*; 276: 26349-26356 (2001)

[58] Massague J., Chen Y.G. Controlling TGF-beta signaling. *Genes Dev* 14, 627-644 (2000)

[59] Yamamoto, S. Akiyama, T. Katagiri, M. Namiki, T. Kurokawa, T. Suda, Smad1 and smad5 act downstream of intracellular signalings of BMP-2 that inhibits myogenic.

[60] Ikeda S., Hachisu R., Yamaguchi A., Gao Y.H., Okano T.. Radiation retards muscle differentiation but does not affect osteoblastic differentiation induced by bone morphogenetic protein-2 in C2C12 myoblasts. *Int J Radiat Biol*; 76: 403-411 (2000)

[61] Esenwein S.A., Sell S., Herr G., Gaissmaier C., Bamberg M., Mollenhoff G., Kusswetter W., G. Muhr, Effects of single-dose versus fractionated irradiation on the suppression of heterotopic bone formation - an animal model-based follow-up study in rats. *Arch Orthop Trauma Surg* 120, 575-581 (2000).

[62] Wiseman, P. W., Hoddelius, P., Petersen, N. O., Magnusson, K. E., (1997). Aggregation of PDGF-beta receptors in human skin fibroblasts: characterization by image correlation spectroscopy (ICS). *FEBS* 401:43-8.

[63] Brown, C. M., Petersen, N. O., (1998). An image correlation analysis of the distribution of clathrin associated adaptor protein (AP-2) at the plasma membrane. *J. Cell Sci.* 111:271-81.

[64] Brown, C. M., PhD Thesis, The University of Western Ontario, London, Canada, 1998.

[65] Brown, C. M., Roth, M. G., Henis, Y. I., Petersen, N. O., (1999). An internalization-competent influenza hemagglutinin mutant causes the redistribution of AP-2 to existing coated pits and is colocalized with AP-2 inclathrin free clusters. *Biochemistry* 38:15166-73.

[66] Srivastava M., Petersen, N. O., (1998). Diffusion of transferrin receptor clusters. *Biophys. Chem.* 75:201-11.

[67] Srivastava M., PhD Thesis, The University of Western Ontario, London, Canada, 1998.

[68] Petersen N. O. (2001) *Fluorescence correlation spectroscopy*. R. Riegler and Elser Springer Verlag.

[69] Nohe A. and Petersen N. O. (2002). Image Correlation Spectroscopy: Exploring Intramolecular Interactions. *Biophotonics* 9: 39-52.

[70] Nohe A. and Petersen N. O. (2003). Analysing the Protein-Protein Interactions in Cell Membranes. *Bioessays*. Accepted.

In: Focus on Cellular Signalling Research
Editor: Dorothy T. Leeds, pp. 73-107
ISBN 1-59454-619-3

Chapter IV

Subcellular Movement of Signalling Molecules. How and Why?

Graciela Piwien Pilipuk and Mario D. Galigniana*
Instituto de Investigaciones Bioquímicas-Fundación Instituto Leloir
Facultad de Ciencias Exactas y Naturales, Universidad de Buenos Aires

Abstract

The biological function of proteins is determined by their cellular localization and subsequent interactions with other factors within a given subcellular compartment. Therefore, it is critical to understand how proteins move to and from the sites where they exert biological effects and how this mechanism of movement is regulated. This concept becomes particularly interesting when soluble signalling factors such as STATs, p53, NFkB, MAPKs, C/EBPs, steroid receptors or cyclins are involved. Soluble proteins are not confined to the cytoplasm or the nucleus in a static manner, but they shuttle dynamically between subcellular compartments regardless of where they are primarily localized under certain biological conditions. Consistent with this concept —and not surprisingly—protein mistargeting leads to a number of pathologies. Ideally, most of these pathologies could be attenuated or even prevented if we were able to regulate the subcellular localization of those mistargeted proteins; i.e. by interfering with the molecular machinery for protein movement and/or by regulating the function of anchoring factors of a given compartment. Nowadays, we know no more than the basics about regulatory mechanisms for protein anchoring, so we still have more questions and doubts than answers and certainties. On the other hand, the molecular mechanism by which soluble proteins move in the cell is an unsolved biological conundrum. In this regard, movement has always been assumed to occur in a stochastic manner by simple diffusion. This oversimplified model has been accepted as the driving mechanism for protein movement in both the cytoplasm and the nucleus. Although heuristic, this notion contradicts the concept of efficiency for protein targeting and, even more importantly, it

* Instituto de Investigaciones Bioquímicas-Fundación Instituto Leloir, Facultad de Ciencias Exactas y Naturales, Universidad de Buenos Aires, Av. Patricias Argentinas 435, Buenos Aires C1405BWE, Argentina. Tel: +54 (11) 5238-7500 –ext. 3308; Fax: +54 (11) 5238-7501; e-mail: mgali@leloir.org.ar

collides against the concepts of cellular compartmentalization and specificity of action of signalling proteins. In this chapter we describe a novel model in which the cytoplasmic movement of some members of the nuclear receptor superfamily is regulated by their association with high molecular weight immunophilins. Based on their subnuclear distribution in different conditions, we also reformulate some classical concepts about the functional regulation of archetype hormone-regulated transcription factors such as steroid receptors and C/EBPs, and postulate the existence of a non-random mechanism for intranuclear trafficking of proteins.

Introduction

Over the past few years, many fields of biological research have converged in the study of the subcellular distribution of signalling proteins. Protein transport is a fundamental mechanism for regulating not only protein localization, but also protein function. Thus, it is hardly surprising that several pathologies are related to mislocalization and altered function of a variety of proteins, which may lead to cell death, cell proliferation, or initiation and progression of cancer [1-8]. Nonetheless, a major unsolved problem in Cell Biology is how those non-vesicle-associated signalling molecules move through the cytoplasm and the nucleus to their sites of action.

It is currently accepted that soluble proteins are not confined to the cytoplasm or the nucleus in a static manner, but are capable of shuttling dynamically through the nuclear pore [5,9-11]. Moreover, the reorganization of nuclear proteins is an essential step for them to acquire certain functions and/or repress others. The same unsolved problems described for the molecular mechanism responsible for cytoplasmic movement of soluble proteins are also valid for the intranuclear movement of nuclear factors. Actually, our knowledge about the latter mechanism is even more limited.

Protein transport across the nuclear envelope involves sequential steps: retrograde movement through the cytoplasm, recognition by nuclear import proteins, docking to the nuclear pore, translocation across the pore, movement through the nuclear compartment, anchorage to and release from the nuclear sites of action, recognition by nuclear export proteins, anterograde translocation across the nuclear pore, and anterograde cytoplasmic movement. Several studies have been conducted to elucidate import and export mechanisms through the nuclear pore and, even though there is an uncountable number of questions that still deserve an answer, the information currently available in this field overwhelms the information we have on the molecular mechanism/s of protein movement within the cytoplasm and nucleus (see [11,12] and references therein).

One possible answer to our question involves the possibility that soluble proteins just move in a stochastic manner by simple diffusion, such that random collisions occurs between soluble proteins and cell structures. Consequently, after an effective collision, signalling proteins become trapped at their sites of action by protein-protein or protein-nucleic acid interactions. If the movement mechanism were this, it would be difficult to explain how each protein exerts specific effects when a given cascade is activated since the protein responsible for triggering the process would freely spread throughout one or more cell compartments.

A priori, a mechanism of movement solely based on free diffusion of soluble signalling proteins also appears to collide with a basic biological concept, i.e., the fact that the work is highly compartmentalized in the cell. Inasmuch as proteins normally occupy the entire cell compartment upon their activation, additional mechanisms, which may include specific protein-protein interactions, must regulate protein movement. An alternative regulatory mechanism involves the time-frame of a given event, such that when one cascade is activated, other cascades that are also triggered by the same signal became immediately repressed or modulated. If this is the case, it raises the question of how a given signalling cascade and not others is activated in first term. One possible explanation is the putative vicinity of the components of the activating machinery, in which case a simple diffusion-based mechanism may be compatible with some degree of specificity within a given time-dependent frame. Cytoskeletal filamentous macromolecules have net negative charge [13] and are able to interact with arginine and lysine groups of signalling proteins. Thus, electrostatic interactions could restrict the motion of soluble solutes to the vicinity of filaments and provide the conditions for a 'directed' diffusion along cytoskeleton constraining the molecules into more focused trajectories. Furthermore, the speed of signal transduction could be controlled by the cell by changing the cytoskeleton structure.

However, the crowded intracellular environment may compromise the efficiency of this free diffusion, which becomes highly anomalous for large structures whose size is equivalent to oligomeric macromolecules like, for example, steroid receptor•hsp90•immunophilin heterocomplexes [14]. It has also been determined that diffusion in membrane-adjacent cytoplasm is significantly slower than that measured in bulk cytoplasm due to the high density of proteins near the cell membrane [15]. The crowded state of cell interior was visualized by cryoelectron tomography, and the images support the view that cytoplasm is filled with large and highly packed assembles of filaments and macromolecules forming functional complexes rather than freely diffusing and colliding soluble complexes (reviewed in [16]). Moreover, due to the large size and extreme polarity of neurons, it is accepted that in a specialized cytoplasmic structure such as the axon, random diffusion would not permit delivery of protein solutes over long distances [17-19]. In fact, distances can reach one metre or more in humans and large animals, and axonal volumes can exceed the volume of the cell body by three orders of magnitude.

An alternative and more directed mechanism for protein movement is that in which solutes utilize a movement machinery to traverse cell compartments, so movement would be likely to involve molecular motors and cytoskeletal or nucleoskeletal tracts in an active manner. This option implies the concept that specific interactions would be required to determine the direction of signal protein movement. We will address this subject below.

If cytoplasmic motion by simple diffusion is largely distorted for soluble proteins in the cytoplasm, the situation should be even more dramatic in a more compacted and dense compartment such as the nucleus. Studies on subnuclear localization of catalytic and regulatory proteins within the nucleus are increasing exponentially. The common and universal finding is that signalling proteins are not homogeneously distributed but enriched at steady-state levels in discrete structures. In short, the well-established concept of compartmentalization associated to the cytoplasm is also valid for the nucleus. Importantly, chromatin changes dynamically and the subnuclear organization of factors is efficiently and rapidly achieved upon cell activation with specific stimuli. The question is similar to that we asked for cytoplasmic movement, how soluble factor mobility occurs in the nucleus?

The Steroid Receptor•hsp90•IMM Heterocomplex

Steroid receptors (SRs) are an excellent model for studying protein movement because their subcellular localization can be placed under the control of the investigator by adding or washing out the hormone. In the absence of ligand, some SRs are primarily cytoplasmic, i.e. the glucocorticoid receptor (GR) [20], mineralocorticoid receptor (MR) [21], androgen receptor (AR) [22], dioxin receptor (AhR) [23], vitamin D receptor (DR) [24], etc. Upon ligand binding, these receptors move rapidly (minutes) towards the nucleus, whereas they cycle back to the cytoplasm upon ligand withdrawal in a slower manner (hours). In their mature form, SRs exist associated to 90-kDa and 70-kDa heat shock proteins, the small acidic protein p23, and proteins that posses tetratricopeptide repeats (TPR), i.e. sequences of 34 amino acids repeated in tandem (Fig.1, *black crescent*) that are critical for protein-protein interactions [25]. In the SR heterocomplex, the TPR-acceptor site of hsp90 is normally occupied by either high molecular weight immunophilin (IMM) FKBP52, FKBP51, CyP40 or PP5. IMMs are a family of intracellular receptors for immunosuppressant drugs also characterized for having peptidylprolyl isomerase (PPIase) enzymatic activity, which directs *cis-trans* isomerization of peptidylprolyl bonds [26]. The oligomeric structure of untransformed SRs is basically composed (among other proteins) by one molecule of SR, a dimer of hsp90 and one high molecular weight IMM, and is also found in primarily nuclear untransformed receptors such as the progesterone receptor (PR) or the estrogen receptor (ER). It should be emphasized that both cytoplasmic and nuclear receptors, are not confined to one particular compartment in a static manner, but they undergo an active and dynamic nucleocytoplasmic shuttling.

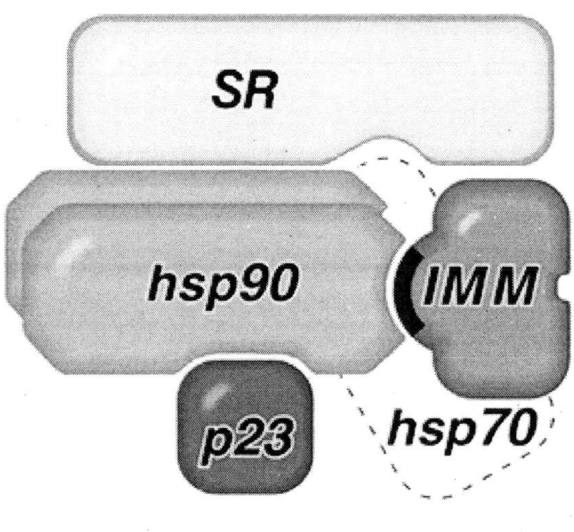

Fig.1. Mature steroid receptor heterocomplex.

A classical model accepted *à bouche ouverte* for almost two decades posits that the ligand binding-dependent dissociation of the hsp90-based heterocomplex (a process frequently referred to as 'transformation') is a *sine qua non* requirement for the nuclear translocation of SRs [27-29]. This model was based on the assumption that the chaperone complex anchors SR to the cytoplasm impairing its nuclear translocation. On the other hand, the constitutive nuclear localization of ER and PR was assigned to the fact that their nuclear localization signals (NLS) are more protruding (and consequently more exposed) than the NLS of primarily cytoplasmic SRs. This model was challenged in some early publications where it was postulated that both hsp90 [30-32] and the hsp90-binding IMMs might be involved in protein trafficking [33], and by the fact that the GR co-localizes with microtubules [34]. However, the transformation model, although unproved, finally prevailed during the following years.

A very important finding was the discovery that cytoplasmic dynein co-immunoprecipitates with FKBP52 [35]. Dyneins are molecular motors that generate force towards the minus end of microtubules and are related to the retrograde movement of vesicles [36]. Cargo attachment occurs via the dynein intermediate chain, whereas the ATP-hydrolytic domain responsible for the motor function is located in the heavy chains. The actual microtubule-binding site is a small globular unit that protrudes from heavy chains. Then, it became clear that the dynein-IMM interaction involves the PPIase domain of the IMM [20], this property appearing to be a common feature for most high molecular weight IMMs associated to SRs [37]. Importantly, at least two of the most abundant IMMs found in SR•hsp90 complexes, FKBP52 and PP5, co-localize with microtubules [37-39].

These observations implied that SRs may move towards the cell nucleus utilizing an active transport system that would require hsp90, IMMs, dynein motor proteins and cytoskeletal tracts. If correct, one critical corollary of this hypothesis is that the hsp90•IMM complex associated to the untransformed receptor should not dissociate upon ligand binding because it would be required for the retrograde movement of the ligand-receptor complex. The experiment described in Fig.2 clearly demonstrates that, in contrast to what the unproven classical model has always stated, transformation is not an early event. Fig. 2-A shows that upon steroid binding, the MR rapidly translocates to the nucleus in renal collecting duct cells, so hundred percent of the MR population becomes entirely nuclear after ten minutes of incubation with aldosterone. When the [^{3}H]aldosterone-MR complexes in the nuclear compartment are analyzed by a continuous sucrose gradient (Fig. 2-B), more than 70% of the MR population still co-migrates with the 9.4 S, untransformed MR form, even ten minutes after the steroid was added to the medium, i.e., a time when the MR is totally nuclear (Fig.2-A). This observation suggests that hsp90 is not immediately dissociated from SRs upon ligand binding. This notion is evidenced in a more convincing manner by resolving the gradient after a preincubation with an IgM antibody against hsp90, which switches the 9.4 S peak to 11.4 S (Fig.2-B). It should be pointed out that after 15 minutes with hormone, the 9.4 S peak is greatly reduced in favour of the transformed, 5.4 S MR form, whereas after 20 min in the presence of hormone, hundred percent of the MR population is fully transformed and tightly bound to chromatin (data not shown).

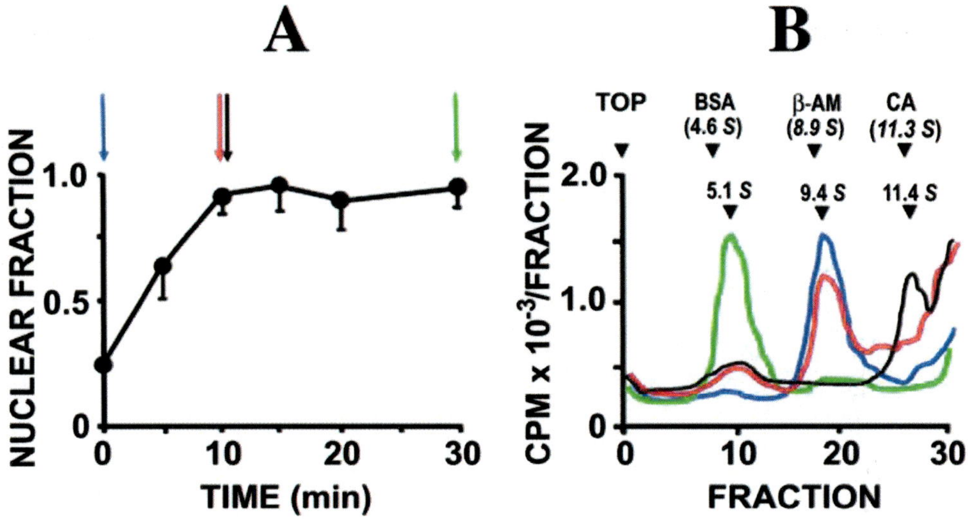

Fig.2. Transformation of SRs is not an early event. *A.* Nuclear translocation rate of the MR after treatment of renal duct cells with 10 nM aldosterone. *B.* Sucrose gradient of [^{3}H]aldosterone-MR complexes in whole lysates before adding steroid (*dotted*); gradients of nuclear fractions after 10 min (*grey*) and 30 min (*dashed*) with steroid. The profile of cells treated for 10 min with steroid whose fractions were preincubated with an IgM antibody against hsp90 is shown in *black.* Calibration standards (bovine serum albumin, ß-amylase and catalase) are shown on the top. Arrows in panel A mark the incubation times of those gradient profiles shown in panel B.

In a more conclusive set of experiments (see Fig.3), we demonstrated that the cytoplasmic-nuclear movement of hsp90•IMM-chaperoned factors such as the GR [20,40] and p53 [41] is impaired or blocked by hsp90 inhibitors (i.e. geldanamycin or radicicol), by overexpression of the PPIase domain of FKBP52 that prevents dynein binding, by saturation of the hsp90•IMM binding site with the TPR peptide, and by overexpression of the dynactin complex subunit, p50/dynamitin, which disrupts the dynein-dynactin complex and dissociates cargoes from dynein. Thus, there is strong evidence that validates a model in which the retrograde movement of some soluble factors takes place in an active manner via cytoskeletal tracts, the hsp90•IMM complex being the bridge between the cargo and the motor protein. Fig.3 shows an illustration for the molecular machinery responsible for the retrograde movement of the GR and where some inhibitory agents uncouple the system.

The interaction between the hsp90-binding IMMs with dynein motor proteins seems to be conserved in nature. It has recently been described that plants possess two hsp90-binding IMMs named *w*FKBP73 and *w*FKBP77, that show high homology with mammalian FKBP52 [42]. Interestingly, mGR•*w*hsp90•*w*FKBPs•dyenin heterocomplexes can be assembled in wheat germ lysate [43], and the resultant mGR-chaperone heterocomplex binds steroid with as high affinity as complexes assembled with mammalian proteins. An excess of recombinant TPR peptide in the lysate prevents binding of both wheat IMMs and dynein to the mGR•*w*hsp90 complex, and the PPIase domain of FKBP52 prevents dynein binding to the GR•*w*hsp90•*w*IMM complex. In spite of the fact that the biological function of high molecular weight IMMs in plants remains poorly understood, retention of the entire heterocomplex assembly machinery in plants implies that this complex plays a cardinal role in the biology of the eukaryotic cell, most likely protein trafficking via motor proteins. It is important to note that glucocorticoid-dependent gene expression has been shown in tobacco cells co-transfected with rat GR and an steroid-responsive reporter [44]. Thus, all the conditions for ligand-regulated movement of SRs described in mammalian cells also exist in plants.

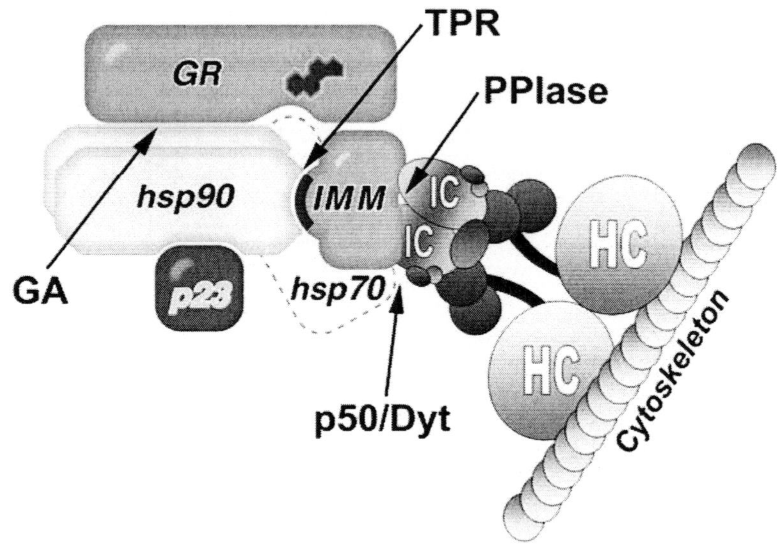

Fig. 3. The GR movement molecular machinery and agents for selective uncoupling of the system. Arrows show the sites of uncoupling by hsp90 disrupting agents such as geldanamycin (*GA*), by the TPR domain fragment of PP5, by the PPIase domain fragment of FKBP52, and by p50/dynamitin (*Dyt*). Immunophilin TPR domain (*black crescent*). Dynein heavy chains (*HC*), intermediate chains (*IC*).

SRs are not the only nuclear factors actively transported to the nucleus by motor proteins. The tumor suppressor p53 was also found to co-localize with microtubules and is associated to dynein [41,45]. As has been shown for SRs, p53 shuttles dynamically between the cytoplasm and nucleus [46,47], and one mechanism of inactivation of p53 is its exclusion from the nuclear compartment [48]. Like unliganded SRs, some p53 mutants are retained in the cytoplasm bound to the same hsp90/hsp70-chaperone system described for SRs [49]. This observation prompted us to study whether the linkage between dynein and the p53-heat-shock protein heterocomplex is a high molecular weight IMM, and FKBP52, PP5 and CyP-40 were found associated to the complex [41]. This association is disrupted by hsp90 inhibitors, the TPR peptide of PP5 or the PPIase peptide of FKBP52 (see Fig.3). More importantly, the cytoplasmic trafficking of p53 towards the cell nucleus was also abrogated in intact cells treated with hsp90-disrupting agents and/or by overexpressing the IMM peptides, indicating that the mechanism of transport is not exclusive for SRs.

Fig.4 shows the comparative nuclear translocation rates for the GR and p53. Clearly, the overexpression of rabbit FKBP52 PPIase domain cannot entirely abolish GR movement towards the nucleus, so the receptor becomes nuclear 50-60 min after hormone binding. In contrast, the overexpression of the PPIase peptide totally abrogates p53 movement. An identical inhibitory effect as that shown for the PPIase domain was also observed when the dynactin subunit, p50/dynamitin, was overexpressed (data not shown). Taken together, these observations demonstrate that, even though the basic machinery for retrograde movement seems to be the same for both factors, the regulation of the nuclear import is governed by additional, nuclear-factor specific mechanisms.

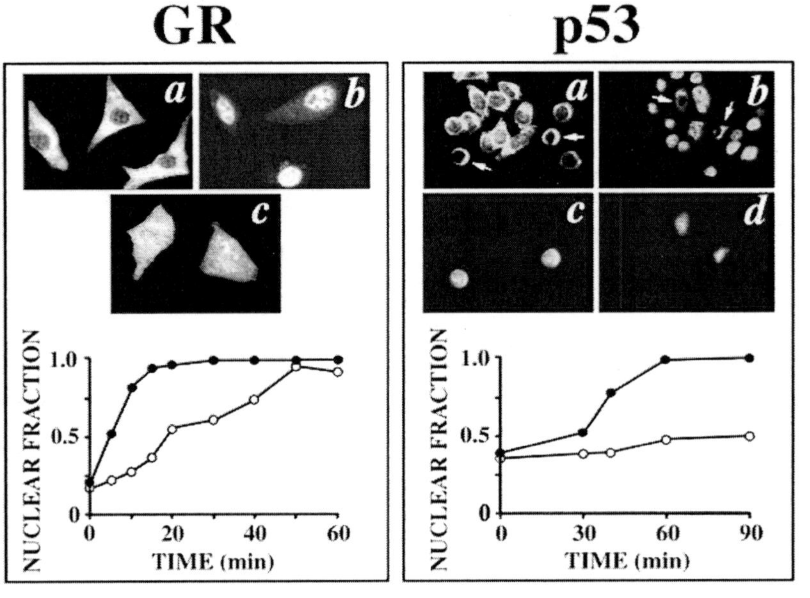

Fig.4. Overexpression of the PPIase domain peptide of FKBP52 inhibits GR and p53 nuclear translocation. *GR*. NIH-3T3 fibroblasts were transfected with GFP-GR and, after 48 hours in a steroid-free medium, dexamethasone was added to the medium at zero time. Panels are: *a,* control without steroid; *b,* incubation with steroid for 20 min; *c,* cells co-transfected with the PPIase domain of FKBP52 and treated with steroid for 20 min. *p53*. HT29-tsp53 colon cancer cells stably transfected with a temperature-sensitive mutant of mouse p53 were grown at 39°C (panel *a*), and p53 translocation to the nucleus was triggered by shifting the temperature to 32°C for 60 min (panel *b*). Arrows show the green cells co-transfected with the PPIase domain of FKBP52 and GFP (panels *c* and *d*). Cells were fixed after 1 h and permeabilized in cold methanol and then immunostained with the Ab-4 monoclonal IgG against mouse p53, followed by incubating with rhodamine-conjugated goat anti-mouse IgG. The nuclear translocation rates for the GR and p53 are shown on the bottom of both panels, Black circles are the control cells and white circles are cells transfected with the PPIase domain of FKBP52.

Dynein does not exhibit motor protein activity by itself, but it requires forming a large protein assembly through its intermediate chains with a unique activator, dynactin. In turn, dynactin is not a single protein but a multimeric complex [50]. It has been reported that the overexpression of p50/dynamitin, one of the various subunits of the dynactin complex, inhibits dynein function by dissociating the motor protein from its cargoes [51,52]. These observations performed on vesicle transport may explain the inhibitory effect observed by overexpressing p50/dynamitin on the retrograde movement of both the GR and p53. However, the real linkage between vesicular cargoes and the dynein-dynactin complex remains unknown. Inasmuch as dynein binds to high molecular weight IMMs, it is entirely possible that IMMs can interact with dynactin and it is p50/dynamitin itself the component of the dynactin complex that interacts with the PPIase domain of the IMM. Fig.5 shows that there is a direct interaction between immunopurified p50/dynamitin and purified FKBP52, this association being disrupted by the PPIase domain of the IMM. Similar results were obtained with PP5 (not shown).

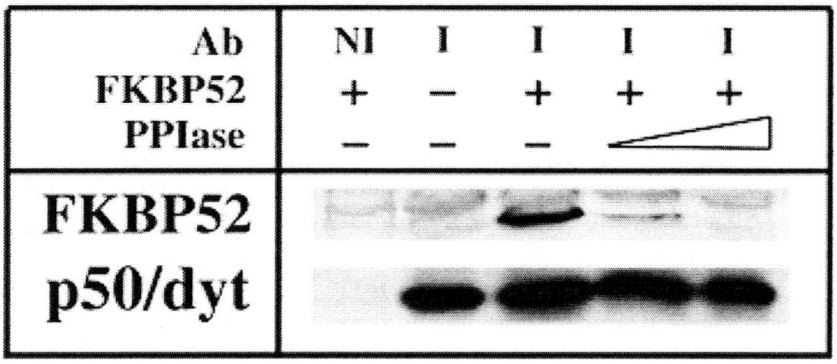

Fig.5. Direct binding of FKBP52 to p50/dynamitin. Cytosols of 293-T cells transfected with *myc*-tagged p50/dynamitin (*p50/dyt*) were immunoadsorbed with anti-*myc* antibody (*I*) or a non-immune antibody (*NI*). Immunopellets were stripped with RIPA buffer and incubated with purified FKBP52 in the absence or presence of purified PPIase domain of FKBP52.

Recruitment of IMMs and Dynein to the SR•hsp90 Complex upon Ligand Binding

Steroids can be found in plants and fungi, but not their cognate receptors as we know them in metazoans. This suggests an independent gain of ligand binding capacity during evolution. On the other hand, it is interesting to point out that, once a given receptor has acquired the capacity to bind a particular type of ligand, it is not greatly modified during evolution, perhaps due to the relative advantage provided by ligand-dependent activation of the receptor. Given the complexity of the metabolic pathways for ligand synthesis, adaptation of an ancestral receptor for the binding of such molecules seems more reasonable than adaptation of a whole biosynthetic pathway of the ligand to the receptor. Clearly, the endocrine system is an issue of evolution that has prompted today biochemists to revise the old hypothesis that the hormone and its receptor could have been pre-existing structures that evolved independently. It is now clear that their interaction may necessarily be the result of evolution itself, and regardless of their origin, both ligand and receptor always function as a unique functional unit [53]. Within the boundaries of this conceptual context, it is likely that the switch responsible for turning on a given signalling cascade, the hormone, may provide particular properties to the cognate receptor to which it is bound. If this is the case, even when the same receptor is considered, each ligand-receptor complex constitutes a differential structural and functional unit.

This concept may explain the versatility and pleiotropism exhibited by certain hormones that exert their biological effects via a common signalling cascade or receptor activation. For example, the potent agonists cortisol and dexamethasone share the GR transformation pathway with the synthetic agonist ΔHOP [10,54]. Thus, ΔHOP efficiently triggers typical glucocorticoid effects such as the induction of hepatocyte TAT activity, anti-inflammatory action, inhibition of nucleotide incorporation into thymocytes, favours glycogen storage in liver, C/EBPδ induction during preadipocyte differentiation, etc. Unlike classical agonists, ΔHOP fails to exhibit other classical glucocorticoid affects, such as down-regulation of the hypothalamus-pituitary-adrenal axis, thymus involution or endometrial cell proliferation. Interestingly, the ΔHOP-GR nuclear import rate is slower than that exhibited by dexamethasone-GR complexes, whereas an accelerated nuclear export for each ligand-GR complex is evidenced in digitonin-permeabilized cells [10]. Limited trypsinization of dexamethasone-GR complex yielded different proteolytic fragments than those obtained with ΔHOP-GR complexes, which is consistent with a differential conformational change generated by each ligand upon binding to the same GR into the same cell context.

A similar property can also be seen for the synthetic mineralocorticoid 11,19-oxidoprogesterone (11-OP) bound to the MR when it is compared to the physiologic agonist aldosterone [55]. A controlled proteolysis of unliganded MR α-chymotrypsin yielded an almost complete proteolytic degradation of the MR (Fig.6-A). However, the aldosterone-bound MR complex still shows a 34-kDa band as a predominant intermediate proteolytic product. Similar results were also obtained with the agonists 11-deoxycorticosterone and corticosterone. Interestingly, the synthetic agonist 11-OP showed and intermediate pattern of degradation between free and aldosterone-bound MR, and only a partial degradation of the 34-kDa fragment was observed.

Fig.6-B shows that the nuclear translocation rate of 11-OP-MR complexes is at least twice slower ($t_{0.5} \approx$ 12-13 min) than that measured for complexes with the corresponding natural agonist ($t_{0.5} \approx$ 5 min). This leads to ask whether the trafficking machinery undergoes any quali-q7uantitative modification upon binding of each steroid. It is reasonable to think that changes in SR conformation can favour the recruitment of certain proteins whereas others are excluded. To answer this question, the MR was immunoprecipitated after incubating the cells on ice with either aldosterone or 11-OP, so steroid binding takes place but not nuclear translocation, which is a temperature- and energy-dependent process. The proteins co-immunoadsorbed with the MR were resolved by Western blotting. Results are shown in Fig.6-C.

Aldosterone binding greatly favours the recruitment of both IMMs FKBP52 and PP5, as well as the motor protein dynein. On the other hand, FKBP51 is dissociated from the complex upon aldosterone binding, presumably due to the fact that the recruited IMMs replace FKBP51 in the unique TPR binding site existing per hsp90 dimer [56]. It is important to emphasize that FKBP51 does not bind dynein and its presence in the SR complex has an inhibitory effect on ligand binding capacity [57,58]. Although to a lesser extent, 11-OP binding also recruits IMMs and dynein, and some FKBP51 can still be recovered in the complex. A lower efficiency for recruiting FKBP52 and dynein cannot be assigned to a smaller amount of immunoprecipitated receptor or hsp90 in cells treated with 11-OP. Therefore, the nature of the ligand bound to the MR differentially influences the conformational changes of the receptor, which in turn, affects the quantity and nature of factors recruited to the SR heterocomplex. A direct consequence of this differential recruitment of IMMs and dynein is reflected in the nuclear translocation rates measured for aldosterone-MR and 11-OP-MR complexes. Perhaps due to the same reason, the nuclear pattern of the 11-OP-MR complex exhibits a different pattern to the aldosterone-bound receptor, which is more concentrated on specific granular foci (Fig.6-D). At this point of the studies, it is still uncertain whether the number and/or quality of proteins recruited by the MR in the nucleus parallels the property observed for the cytoplasmic form of the receptor. Nonetheless, preliminary studies where the co-activator CBP/p300 was co-transfected have shown a more exacerbated potentiation in the activity of a gene-reporter for 11-OP-MR complexes than for aldosterone-MR complexes.

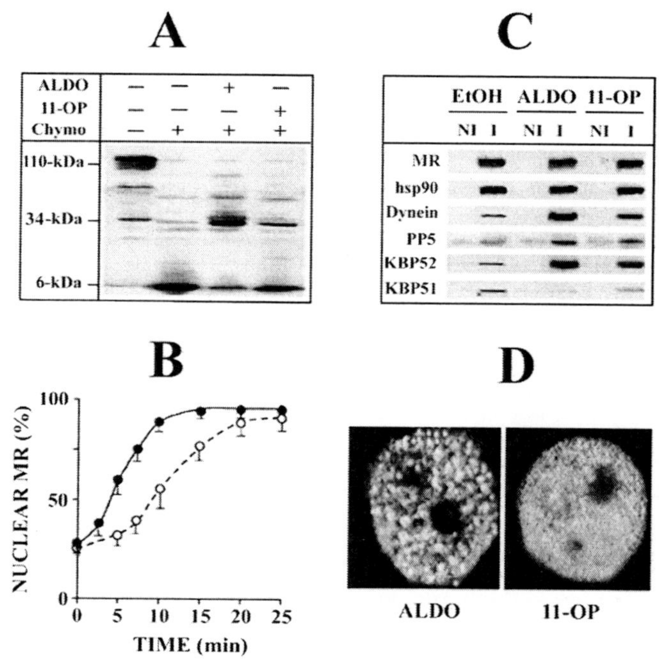

Fig.6. Ligand-dependent conformation of the MR. *A*. Rat MR was translated in reticulocyte lysate in the presence of [^{35}S]Met, immunopurified, and incubated with 0.1 μM aldosterone (*ALDO*) or 1 μM 11,19-oxidoprogesterone (*11-OP*). Then, steroid-receptor complexes were partially digested with chymotrypsin (*Chymo*). *B*. E82.A3 fibroblasts transfected with *myc*-tagged rat MR were incubated on ice for 1 h with vehicle (EtOH), 0.1 μM ALDO or 1 μM 11-OP. The MR was immunoprecipitated with an anti-*myc* antibody and the associated proteins were resolved by Western blotting. *C*. Nuclear translocation rate of the MR in cells incubated with aldosterone (-•-) or 11-OP (-○-). *D*. Nuclear distribution of the MR after 1 h incubation with steroid.

Nuclear Import

Unlike the passage of proteins into mitochondria and other organelles, where proteins are unfolded to pass, proteins pass through the nuclear pore intact. The nuclear pore complex (NPC) is a multi-protein structure that spars the double membrane of the nuclear envelope and functions as both a selective barrier and a conduit for the exchange of molecules between the nucleus and the cytoplasm [59-63]. The NPC is composed of ~50 different polypeptides called nucleoporins (Nups), many of which are characterized by the presence of multiple copies of Phe-Gly motifs (or FG-repeats), which may serve as binding sites during the transport process [64]. The major frame-work of the NPC consists of eight spokes flanked by nuclear and cytoplasmic rings that surround a central channel structure. Extending outward from the ring-spoke assembly, there are 35-50-nm-long cytoplasmic fibrils and 50-100-nm-long nuclear fibrils that are joined in a basket-like structure.

Although small molecules may diffuse between cytoplasm and nucleus, molecules greater than 40-50-kDa are transported actively along the central channel of the NPC by signal- and energy-dependent mechanisms. Nuclear protein import can be separated experimentally into two steps, an initial binding to the cytoplasmic face of the NPC, which requires the presence of an NLS but not metabolic energy, and the translocation step, which is reversibly arrested by metabolic inhibitors, low temperature or by wheat germ agglutinin [65-67].

Most signal-dependent nuclear transport is mediated by nucleocytoplasmic shuttling receptors of the importin/karyopherin family, which bind to cargoes in the cytoplasm via their nuclear localization signal (NLS). Proteins containing a NLS first attach to soluble carriers of the importin-ß family, either directly of via an adapter such as importin-α. So far, nearly twenty members of the importin-ß family have been discovered, some of which directly recognize NLS motifs independently of importin-α [11,68]. The importin-cargo complex docks at the cytoplasmic face of the NPC before translocating through the central channel into the nucleus. The interaction of carriers with their cargo is orchestrated by the Ras-family GTPase, Ran, which possesses intrinsic nucleotide interchange and hydrolysis capacity. Ran-GTP is an essential cofactor for the nuclear delivery of NLS-containing cargoes. Thus, Ran-GTP binding to importin-ß results in the dissociation of cargo proteins, and because Ran-GTP is highly concentrated in the nuclear compartment, proteins are immediately delivered upon their arrival to the nucleoplasm. The carrier is then recycled back through the NPC to the cytoplasm, where RanGAP promotes RanGTP hydrolysis and dissociates the complex. Therefore, the GDP-bound form of the carrier predominates in the cytoplasm. This asymmetry in the distribution of GDP-bound and GTP-bound forms of Ran is due to the nuclear compartmentalization of the Ran guanine nucleotide exchange factor, RCC1 (RanGEF), and the cytoplasmic localization of the Ran GTPase-activating protein, RanGAP.

Interestingly, besides its pivotal role in the above-described mechanism for the nuclear import of cargo proteins, importin-ß also acts positively in several interphase roles (recently reviewed in [69]); for example, as a chaperone for highly charged nuclear proteins, as a potential motor adaptor for movement along microtubules, and transducing damage signals from the axons of injured neurons back to the cell body. In contrast, importin-ß plays a negative regulatory role in mitotic spindle assembly, centrosome dynamics, nuclear membrane formation, and nuclear pore assembly. In most of these effects, importin-ß is counteracted by its regulator, Ran-GTP.

There is evidence that transport of relatively small protein cargoes (~70-120 kDa) through the NPC can occur in permeabilized cells in the absence of nucleoside triphosphate hydrolysis [70-72], although it is unclear whether this mechanism occurs in intact cells. On the other hand, efficient transport of larger proteins does require hydrolysable GTP and the small GTPase Ran, which enhances the ability of import complexes to traverse restricted areas of the NPC. It is also important to emphasize that the importin-dependent system does not account for the nuclear import of all proteins. Thus, several signalling molecules such as SMAD [73,74], ß-catenin [75,76], and ERK2 [77,78] can translocate across the NPC by mechanisms that are independent of importin-ß factors. Nonetheless, Nups can still be required, as suggested by the fact that their occlusion with wheat germ agglutinin blocks the nuclear import of all those signal transducers. In short, like importin-ß, those proteins are able to associate directly with Nups [73,75,77].

In view of this scenario, and because it appears that the SR transformation is not an early event, it results interesting to wonder whether the entire SR•hsp90•IMM complex translocates through the NPC. Some of the experiments described here in the previous sections strongly argue in favour of this hypothesis. In effect, the entire 9.5 *S* complex was recovered in the nuclear compartment (Fig. 2), and the hsp90•IMM complex is required for receptor movement on cytoskeletal tracts (Fig. 3). Moreover, IMMs co-localize with the GR in the nucleus (see Fig.9), suggesting that they may remain associated to the receptor all the time. If this is the case, IMMs should bind directly to the receptor independently of hsp90, which must be dissociated at some point of the activation pathway (most likely, in the nucleus). That this is the case is demonstrated by the fact that the incubation of GSH-agarose purified GST-tagged GR with recombinant FKBP52 yields GR-IMM complexes [56]. This complex can be cross-linked with glutaraldheyde (Fig.7-A), a treatment that yields a product whose molecular weight is the exact addition of the respective molecular weight of each protein. Interestingly, Fig.7-A also shows that a 76-kDa proteolytic fragment of the GR is unable to interact with the IMM. Because it is known that limited proteolysis of the GR generates a 45-kDa fragment [79], and Fig.7-A shows that the proteolytic fragment is bound to GSH-agarose, it must consist of GST plus an ~45-kDa amino-terminal segment of the GR, which does not contain the hormone-binding domain, the hinge region, and most of the DNA-binding domain. Because no cross-linked 76-kDa GST-GR fragment·FKBP52 product is obtained, it is reasonable to predict that FKBP52 binds to one or more of these three missing GR domains. In order to investigate what part of the GR interacts with FKBP52, the binding of purified FKBP52 to two truncated mutants of the GR was assayed. As diagrammed in Fig.7-B, GST-500 is a carboxyl-terminal truncation of the GR lacking the hormone-binding domain and the portion of the hinge region carboxyl-terminal to the nuclear localization signal, NL1, at amino acids 491-498 of the hGR [80,81], whereas GST-465 lacks half of the carboxyl-terminal zinc finger and NL1 as well. As shown in the Western blot of Fig. 7-B, FKBP52 binds to GST-500 (*lane 3*), although the binding was reduced compared with GST-GR (*lane 2*); and there is no binding of FKBP52 to GST-465 (*lane 4*). The C terminus of FKBP52 contains a predicted calmodulin-binding site [81], and FKBP52 binds calmodulin-Sepharose in a calcium-dependent manner [82]. Taken together, these results indicate that FKBP52 binds directly to the GR. The precise region of FKBP52 that binds the GR is not known yet, but it appears to be separate from the regions involved in binding to dynein (the PPIase domain) or hsp90 (the TPR domain) because neither the first fragment nor the second competes for GR binding.

One interesting possibility is that IMMs may be involved in the mechanism of SR passage through the NPC. Consistent with this idea, it is known that cytoplasmic intermediate filaments show continuous tracts that penetrate into the nucleus [83-85] and, on the other hand, FKBP52 interacts with the intermediate filament vimentin [86]. Moreover, FKBP52 co-immunoprecipitates with cargo proteins such as the GR or p53 when they are bound to microtubules [41,87], indicating a direct association between cytoskeletal tracts, IMMs, and cargo proteins. Therefore, it is reasonable to speculate that this structural arrangement may be extended to the nucleus like if it were a continuous highway through which soluble proteins move and/or are attached to be modified (for example, phosphorylation).

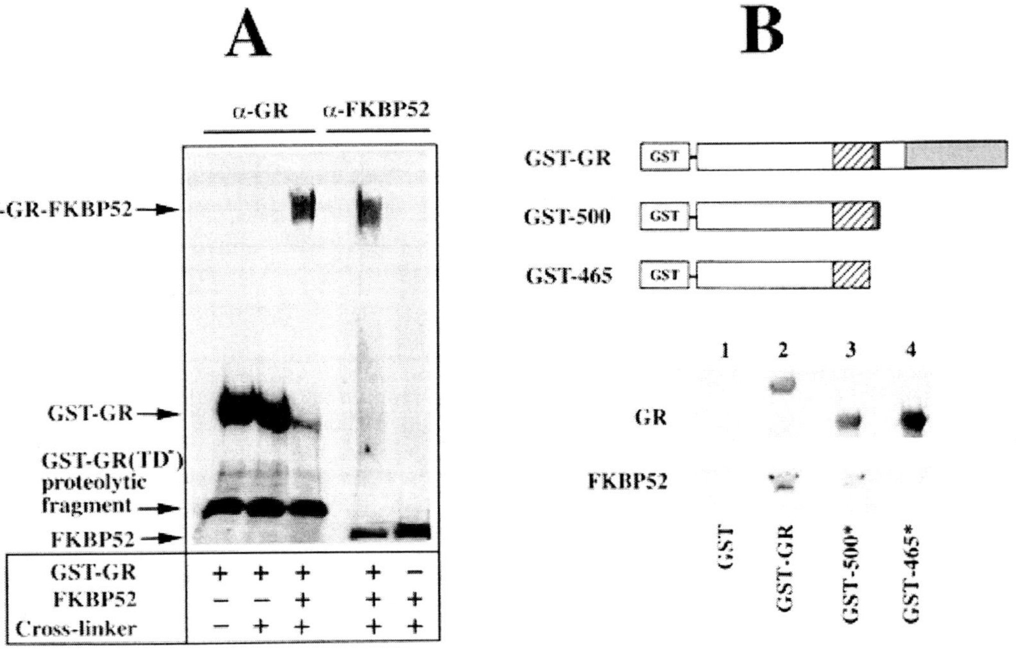

Fig. 7. Direct binding of FKBP52 to the GR. *A*. GST-GR was immobilized on GSH-agarose, stripped of associated proteins by treatment with RIPA buffer for 2 h at 30°C, and incubated on ice for 30 min with purified FKBP52. Complexes were cross-linked with glutaraldheyde and resolved by Western blotting with anti-GR or anti-FKBP52 antibodies. A proteolytic fragment lacking the transactivation domain (TD⁻) is generated during the incubation at 30°C. *B*. Binding of pure FKBP52 to truncated mutants of GR immobilized on GSH-agarose as described in *B*. The GR mutants are diagrammed on the top of the Western blot (the grey box is the LBD and the hatched box is the DBD).

It is also important to emphasize that a functional NLS depends strongly on the protein context. Proteins destined to enter mitochondria, chloroplasts or the lumen of the endoplasmic reticulum contain signal sequences that are often located at the extreme amino terminal [88,89]. It would not appear that, in an analogous fashion, the NLS sequences need to be integrated into a specific region of the protein in order to function. On the other hand, similar basic sequences are also present in a number of non-nuclear proteins, but they are masked. Therefore, while a NLS seems to be required for nuclear localization, its activity is crucially dependent on the protein context. This concept is particularly important for ligand-activated nuclear factors due to the cardinal role played by the ligand on the conformation of the 'activated' receptor. Chaperones and/or IMMs may mask the NLS of primarily cytoplasmic SRs such as the GR or the MR, whereas the conformational change of the receptor generated by ligand binding favours the interaction of the NLS of the receptor with importins. The GR possesses two NLS, NL1 and NL2 [90], NL1 being the best characterized and defined sequence. NL1 is occluded in the unliganded receptor and, upon ligand binding, it becomes accessible to importin-α [91] and an NL1-specific antibody [92].

Two different mechanisms of translocation are evidenced for the GR in Fig.4. Clearly, a more efficient translocation is achieved under physiological conditions than when the SR•hsp90•IMM•motor protein machinery has been disrupted. Although in a less efficient manner (i.e., one order of magnitude longer $t_{0.5}$ for the nuclear translocation rate), GR still moves towards the nucleus. This fact suggests the existence of an alternative mechanism of movement and/or nuclear translocation for the GR, which may be NL1-independent (and perhaps, importin-independent). This speculation is supported by the fact that p53 does not show a similar behaviour when the movement machinery is disrupted; in fact, p53 remains cytoplasmic all the time (Fig.4). One possible explanation for this unequal property exhibited by both nuclear factors may be hypothesized by the presence of the peculiar NL2 of the GR [90], which is totally unrelated to the NLSs present in p53 [48]. The NL2 of the GR resides within a 256 amino acid region that also includes the hormone binding domain. Most importantly, nuclear localization of fusion proteins containing the NL2 region alone is fully hormone-dependent [90]. In this regard, it is important to recall that to date, there are at least 20 different importins described in the literature [93], so it would not be surprising that one of them will be able to translocate the GR via its atypical NL2 signal.

Nuclear Exchange of Steroid Receptors

The eukaryotic nucleus is a highly ordered and compartmentalized organelle. Classical ultrastructural studies have demonstrated that the cell nucleus is an elaborate structure composed of regions of chromatin, which includes the DNA bound with a variety of proteins and attached to a non-chromatin structure, which includes the nuclear matrix [94-98]. Nuclear DNA exists as a hierarchy of compacted chromatin structures, whose basic unit is the nucleosome. The nucleosome consists of 146 base pairs of DNA wrapped around a protein core of histones (H2A, H2B, H3 and H4). The DNA between the nucleosomes is bound by histone H1 or linker histones. Both the linker histones and core histones maintain the higher order folding of chromatin, known as the solenoid, i.e. a 30-nm chromatin filament. In turn, solenoids organize into loop domains with the base of the loop attached to the nuclear matrix. The DNA regions attached to the nuclear matrix are referred to as matrix or scaffold association regions (MARs or SARs, respectively). By definition, the nuclear matrix is the nuclear structure remaining after the nuclease treatment of a salt-extracted nucleus. The nuclear matrix proteins can be classified into two types , those that are common matrix proteins found in all cell lines and tissues (for example, lamins A and C) and those that are specific for certain cell types and differentiation states [94,98].

The development of the EDTA-regressive uranyl acetate staining technique [99] allowed the visualization of nascent RNAs in the form of perichromatin fibrils, which range in diameter from 3 nm to 20 nm [100]. These perichromatin fibrils are also stained by anti-hnRNP proteins [101,102]. Therefore, the nucleus shows fibrilar components that limit an interchromatin space, mostly dominated by perichromatin fibrils. The RNA is visualized as particles associated to protein architecture, but it constitutes a minor component of the interchromatin space. The concept of an intranuclear scaffolding structure was first proposed by Russian investigators in 1948 [103], this structure being named nuclear matrix, karyoskeleton, nucleoskeleton or nuclear scaffold [104]. It shows a tight association with cytoplasmic intermediate filaments, and the crosslinkage of vimentin with genomic DNA fragments, including MARs, point to the participation of intermediate filaments in the nuclear activity [105-108].

The universal finding that proteins are not homogeneously distributed within the nucleus, but rather, are enriched at steady-state levels in discrete foci, has led to explore the hypothesis that there is a coordinated organization and spatio-temporal regulation of nuclear processes. After depleting the nucleus of chromatin, the residual nuclear structure is the so called nuclear matrix, which represents the framework for the organization of the nuclear function [103]. The absence of "intranuclear membranes" has contributed to the early scepticism towards compartmentalization having a role in nuclear regulation. However, we know today that a membrane boundary is not the only mechanism of sequestering molecules within a defined space. Moreover, we can appreciate the dynamic nature of compartments in the nucleus with the advent of GFP technology. The development of receptor fusions with jellyfish fluorescent proteins permitted the examination of receptor localization in living cells, and multiple manuscripts reported in a very short period of time the specific location in foci domains for almost all members of the steroid receptor subfamily [109-114].

The nuclear matrix has been implicated in the actions of SRs on gene expression [115-119], and both chromosome structure and nuclear matrix interaction appear to be important in steroid hormone-mediated transcriptional control. It has been noted that in the absence of ligand, the nuclear localization of SRs is diffuse, whereas the formation of discrete foci is agonist-dependent [10,109-111,120,121]. Therefore, it is possible that SRs that have reached the nuclear compartment move to staging areas after crossing through the NPC, such that they can proceed to the discrete foci upon hormone-dependent transformation.

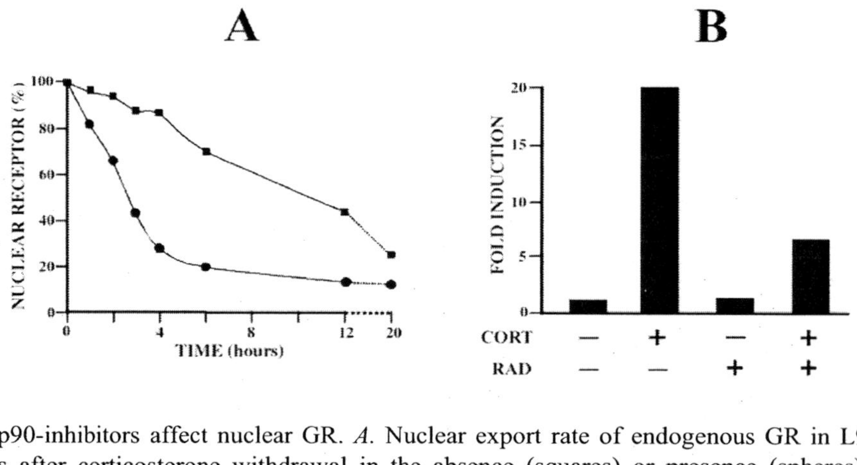

Fig.8. Hsp90-inhibitors affect nuclear GR. *A*. Nuclear export rate of endogenous GR in L929 mouse fibroblasts after corticosterone withdrawal in the absence (squares) or presence (spheres) of 5 μM radicicol. *B*. Induction of luciferase gene reporter activity by 100 nM corticosterone is inhibited by 5 μM radicicol (*RAD*) in WCL2 cells overexperssing constitutively nuclear GR.

There is some evidence that the cycling of SRs into SR•hsp90 complexes may be involved in the steroid-dependent movement of the SR to those discrete foci. As shown in Fig.4, hsp90-disrupting agents impair the nuclear translocation rate of the GR and p53.

However, it appears that cytoplasmic movement is not the only step affected by the inhibitors since they also exhibit effects on nuclear GR. Fig.8 shows an accelerated nuclear export of the GR upon cell treatment with the hsp90 inhibitor radicicol, whereas agonist-dependent transcriptional activity is impaired in radicicol-treated WCL2 cells (a cell line where the GR is constitutively nuclear [122]). These results suggest that hsp90 activity is required for ligand-SR complexes as to promote the proper conformation required for regulating nuclear transfer, nuclear matrix binding and transcriptional activity. In agreement with this hypothesis, it has recently been reported [123] that the overexpression of a nuclear-targeted NLS-fused hsp90 attenuates the GR nuclear export rate and that geldanamycin blocks the recovery of nuclear GR mobility elicited by reticulocyte lysate in digitonin-permeabilized cells [124], which suggests again that the hsp90 complex may be involved in anchoring SRs to nuclear structures. Intriguingly, overexpression of p23 increases the transcriptional activity of the GR and the PR, whereas it decreases the activity of the MR, ER, AR, TR and RAR [125]. In a cell-free system assay, p23 inhibited the transcriptional activation by preformed regulatory complexes, consistent with a direct role of p23 in promoting disassembly of the complexes [126]. Unfortunately, it was not studied whether disassembly of these complexes is ATP-dependent. Inasmuch as GR release from the nuclear matrix has been found to be an energy-dependent step [127], one alternative interpretation for the effect of p23 on transcriptional activity of SRs is that this co-chaperone functions via its known effect on hsp90, which stabilizes the client protein•hsp90 complex [128-131].

Nothing is currently known about how molecules move into the nucleus. In view of the above-described evidence, it is possible that molecular chaperones may participate in various stages of SR movement within the nucleus and facilitate both SR exchange from specific binding sites within chromatin and overall mobility as the receptor moves through the nucleoplasm. There is evidence that some proteins move by simple diffusion in the intranuclear space (recently reviewed in [103,132,133]). Moreover, the existence of the nuclear matrix itself has been questioned [103]. Nevertheless, the experimental foundation of nuclear matrix research lays on the observation by electron microscopy of fibro-granular non-chromatin nuclear structures [134-139], the biochemical isolation of non-chromatin structures [140-143], the organization of chromatin in loop domains attached to non-chromatin structures [144-146], and the discovery of spatially distinct nuclear domains that remain in place after removing soluble proteins [110,141,147-149].

Nuclear protein dynamics has almost exclusively been studied by overexpressing GFP fusion proteins (transcription factors and proteins related to DNA replication). In all cases, a relatively rapid exchange of the protein under study has been measured, and there is a general consensus to state that diffusion is the driving force for nuclear protein movement. A first analysis of those data leads to accept that the conclusions are correct since this is what the measurements are telling us. However, there is a missing point not always considered when those data are evaluated, i.e., the fact that GFP-tagged proteins are overexpressed. Therefore, the interpretation that proteins move by simple diffusion in the nucleus would be meaningful if, and only if, the number of GFP fusion protein molecules plus the number of endogenous molecules is considerably smaller than the number of available binding sites. If the binding sites were saturated, it is expected that the excess of protein must freely diffuse, masking the minimal proportion of attached molecules to their binding sites. As a result, the overall measurement of protein mobility should be more compatible with a diffusion-dependent mechanism than with a biochemically-targeted mechanism similar to that accepted to occur in the cytoplasm.

The possibility that the current model may be the result of a misinterpretation of the experimental data may be considered from various perspectives. First, we know that the nuclear receptor population specifically bound to the cognate site of action is as low as 1 -2% of the total. Thus, in physiological conditions, 98-99% of the total population of an endogenous nuclear factor such as SRs is not associated to hormone response elements. The percentage of specifically bound receptor is even lower if the protein is overexpressed and consequently saturates both the nuclear target sites and staging areas. Therefore, it is expected that this excess of protein freely diffuse in the interchromatin space with the only limitations of its size and charge [150].

On the other hand, movement of splicing factors in the nucleus was found to be dependent on the phosphorylation state and the transcriptional activity of the nucleus, suggesting that the nuclear movement of these factors occurs in a regulated manner rather than at random [151]. The mobility of the splicing factor ASF is 100-times slower than that expected for free diffusion, and is consistent with frequent and transient interactions with relatively immobile nuclear binding sites [152]. Interestingly, in these studies it was also observed ASF-enriched particles budding and moving away from nuclear speckles. These speckles dynamically unfold into a fibre and refold back into a speckle, whereas their positions remain relatively stable within the nuclear space. Other interesting observation was performed with the nucleolar phosphoprotien Nopp140, which shuttles between the nucleolus and the cytoplasm. Shuttling of Nopp140 is likely to proceed on tracks that extend across the nucleoplasm from the dense fibrillar component of the nucleolus to pore complexes [153]. However, the nature of the putative fibrillar tracks as well as the identity of potential motor proteins as we know them in the cytoplasm remains unknown.

In view of the fact that nuclear protein function is often directed to specific subnuclear compartments, a movement system might exist to facilitate the targeted delivery of protein solutes. In this regard, unliganded nuclear SRs such as the PR or the ER are also recovered in complexes with hsp90, hsp70, p23, and high molecular weight IMMs [154], which is also in agreement with the notion that the hsp90•IMM complex required for protein movement in the cytoplasm may also be involved in targeting proteins within the nuclear compartment. Accordingly, indirect immunofluorescence studies demonstrate a clear nuclear co-localization of FKBP52 and PP5 with the dexamethasone-GR complexes (Fig.9). It is interesting to emphasize that these images show the entire nuclear population of the GR co-localizing with the IMM. However, the opposite is not true and speckles of IMM that do not co-localize with receptor are also seen, suggesting that nuclear IMMs play additional roles in the nucleus beyond its association with SRs.

Nuclear Compartmentalization and Relocalization of Proteins

In the nucleus, each chromosome occupies its own distinct region. Chromosomes with low gene density reside at the nuclear periphery, whereas chromosomes with high gene density are located in the nuclear interior [155-157]. While a spatial organization of chromatin in the nucleus is more or less established today, its functional significance remains poorly understood. Transcriptional silencing might occur if a locus were moved to a compact chromatin domain where transcriptional factors might be sterically excluded or where silencing factors are concentrated. On the other hand, transcriptional activation could be facilitated if a locus were relocated to an open and more accessible environment. Thus, RNA transcripts could be released directly into the interchromosomal domains for transportation to the nuclear pores [158]. However, there is little evidence that heterochromatin provides an access barrier to either chromatin proteins or transcription factors. Actually, immunolabelling of nascent BrdU-labelled RNA together with chromosome territory painting in human fibroblasts revealed that RNA transcripts can also be found throughout the interior of chromosome territories [159]. Interestingly, active genes can also be observed outside the chromosome territory [160], adding more evidence that the genome is highly organized.

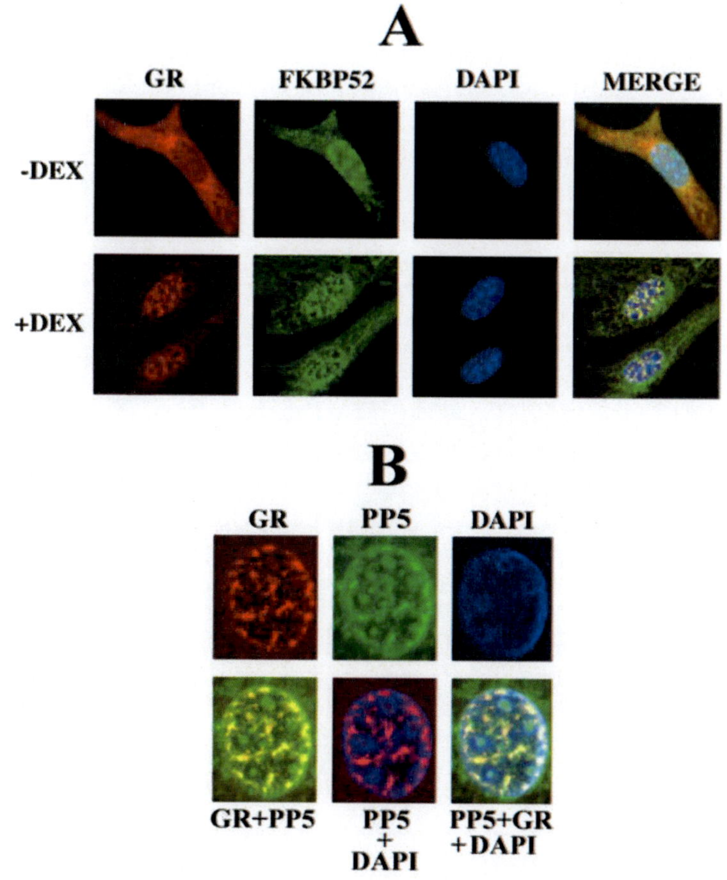

Fig.9. Co-localization of IMMs with the GR. F442A fibroblasts were treated with dexamethasone (DEX) to translocate endogenous GR to the nucleus. Nuclear co-localization of the receptor with FKBP52 (panel *A*) and PP5 (panel *B*) was evidenced by indirect immunofluorescence followed by confocal microscopy.

However, an influence of chromatin structure on the accessibility of transcription factors to their sites of action that is solely based on the concept of steric exclusion seems inconsistent with the biological facts. It is more reasonable to think that the overall spatio-temporal regulation of gene expression would require not only subnuclear organization, but also the relocalization of nuclear factors by a specific mechanism of movement that facilitate and control the assembly of multimolecular transcriptional regulatory complexes. As a consequence, the change in subterritorial positioning of a given locus would be reached in a more regulated manner.

It is interesting to emphasize that nuclear movement of SRs appears to be energy-dependent. Under conditions that lead to more than 95% depletion of cellular ATP levels, the GR remains trapped in the nuclear compartment and does not efflux to the cytoplasm upon hormone withdrawal [127]. Interestingly, the GR remains tightly associated to the RNA-containing nuclear matrix throughout all planes of the nucleus.

We can envision that a lot of molecules and complexes are moving in a crowded environment. Even though we cannot discard that complexes can move by simple diffusion, a directed (ATP-dependent?) movement of molecules to their sites of action may account for a more efficient intranuclear trafficking. Inasmuch as targeting nuclear proteins seems to be regulated in a highly specific manner, it is unlikely that randomized nuclear movement can account solely for such specific and efficient mechanism.

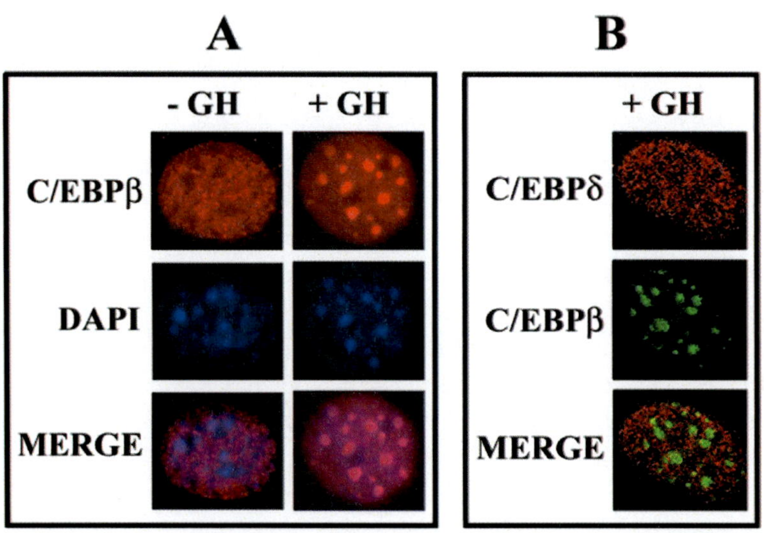

Fig.10. C/EBPß moves to heterochromatin areas upon activation with growth hormone (GH). Nuclear localization of C/EBPß and C/EBPδ was visualized by confocal microscopy upon stimulation with GH. *A.* C/EBPß signal co-localizes with DAPI staining. *B.* In contrast to C/EBPß, C/EBPδ does not modify its nuclear distribution with GH.

Specificity for protein relocalization can be appreciated in the experiment shown in Fig.10. In quiescent 3T3-F442A preadipocytes, the transcription factor C/EBPβ is distributed throughout the nucleoplasm in a diffuse pattern. Several stimuli known to induce transcriptional activity of C/EBPß (GH, insulin, PDGF, IGF, serum) surprisingly promote the redistribution of C/EBPβ to areas of pericentromeric heterochromatin rather than euchromatin. Relocalization of C/EBPβ to heterochromatin seems to depend on MAPK activation since U0126, a MEK inhibitor, prevents its nuclear redistribution [161]. Therefore, it is tempting to speculate that MAPK-dependent phosphorylation of C/EBPβ and/or other protein(s) involved in the nuclear redistribution (movement) of C/EBPβ can be required for targeting this transcription factor to heterochromatic areas. On the other hand, the highly homologous partner C/EBPδ, also inducible by most of those stimuli [162,163], is not affected in the same cells in which C/EBPß is relocated upon hormone stimulation. We still do not know whether this nuclear rearrangement mechanism requires energy as it is the case for SRs, but it seems to be equally affected by hsp90-disrupting agents. This suggests that nuclear chaperones may form part of a more basic regulatory mechanism for protein movement than that proposed for SRs in the cytoplasm and nucleus.

It is important to emphasize that in the case of SRs, their nuclear redistribution upon ligand binding and association to certain nuclear territories appear to be correlated with transcriptional activation. As described previously, we proposed some time ago that the biologic effects of the synthetic glucocorticoid ΔHOP may explain the different GR-dependent response triggered by this ligand compared to the classical agonist dexamethasone [10,54].

This hypothesis was based on a series of properties of the ΔHOP-GR complex, such as a less stable association with nuclear structures, a more diffuse nuclear distribution than the punctuated pattern observed with dexamethasone, a higher solubilization of the GR from nuclear preparations after hypotonic lysis, an accelerated nuclear export rate in permeabilized cells, and by evidencing a different ligand-dependent conformation of the receptor. The hypothesis that differences in nuclear retention of ligand-receptor complexes by differential anchorage to nuclear structures may determine certain biological responses was strengthened when it was described that nuclear mobility correlates inversely with steroid-binding affinity for the receptor [164]. More recently, it was suggested that longer GR residence times are consistently associated with greater transcriptional output [165].

Taken together, these observations raise a new paradigm in which the exchange rate of nuclear factors provides a means to regulate biological effects. It is therefore unlikely that such a sophisticated and cardinal step of the cell life is abandoned to diffusion-dependent, and therefore, stochastic mechanisms rather than a more controlled and regulated machinery of movement.

Conclusion

Inherent to an understanding of cellular signalling is the ultimate comprehension of how signalling proteins travel to their sites of action. Initially, interest focused on the nature of the signals that target movement of protein to specific cellular compartments regardless of the intimate molecular mechanism that commands and regulates the movement. More recently, researchers began to focus on this conundrum and become more and more interested in the underlying molecular basis of nucleocytoplasmic shuttling. In this regard, we have discussed here a model in which a molecular heterocomplex basically composed by hsp90, a high molecular weight IMM and dynein motors, is responsible for the cytoplasmic retrograde movement of SRs and p53.

However, an important question still unanswered is what happens with those proteins that undergo nucleocytoplasmic shuttling and are not chaperoned by hsp90. Is the chaperone system absolutely required? The answer is 'no' because neither hsp90-disrupting agents nor IMM drugs affects the movement of all the proteins in the cell. Thus, an alternative mechanism must exist and perhaps, the hsp90•IMM complex is just one of the various still unknown transport systems of the cell. However, one may envisage that even though a soluble protein is not recovered associated to hsp90 or an IMM, it would be possible that transient interactions with those proteins associated to the cytoskeleton and motor proteins may result in a more efficient trafficking of proteins towards the nucleus. Another speculation is that, like in the case of the GR (Fig.7), it might be possible that proteins can directly interact with the IMM•dynein complex without a chaperone linker. We have recently focused our studies on the low molecular weight IMM, CyP-A [166]. CyP-A shows a PPIase domain but not the hsp90-binding TPR domain present in high molecular weight IMMs. However, CyP-A was recovered in native heterocomplexes containing cytoplasmic dynein. These complexes could be reconstituted in a cell-free system, and these assays demonstrated that the PPIase domain fragment of FKBP52, but not the TPR domain of PP5, blocks complex formation. The complex also contains tubulin if cytosol is prepared under microtubule-stabilizing conditions. Moreover, CyP-A co-localizes with microtubules in intact cells, this co-localization being disrupted by overexpression of the PPIase domain fragment of FKBP52. Like the results shown here in Fig.5 for FKBP52, CyP-A also binds directly to p50/dynamitin in a PPIase domain-dependent manner. Thus, it is clear that CyP-A associates *in vitro* and *in vivo* with the dynein/dynactin motor protein complex in the absence of hsp90.

Interestingly, there are several lines of evidence suggesting that HIV-1 uses dynein and the microtubule network to transport its genome to the nucleus [167] and that CyP-A seems to be required for virus infectivity [168-170]. It is therefore tempting to suggest that CyP-A may perform a general function related to the binding of cargo for retrograde movement along microtubules, including viral particle transport.

In addition to the existence of molecular machinery able to transport soluble proteins to the nucleus in a more efficient and directed manner, the final subcellular localization of a given factor depends on the balance of signals able to activate the factor as to target it towards one or another compartment. Additionally, the localization of soluble proteins also depends on the balance between nuclear retention and nuclear export. A good evidence is the observation that cisplatin cross-links nuclear matrix-bound transcription factors to nuclear DNA in breast cancer cells, such that tumour progression is abrogated by inhibiting transcription and replication [171,172].

All three mechanisms (transport machinery, biochemical signals, and nuclear retention) can be targeted with therapeutic ends to promote an unbalanced redistribution of a signalling protein. For example, the nuclear factor NFκB, which is constitutively activated in many cancer cells, has been implicated in tumour development [173], whereas p53 activation promotes cell-cycle arrest and apoptosis, so p53 mislocalisation in the cytoplasm is associated to tumour development [174]. Therefore, in those cases, it would be possible to maintain p53 in the nucleus and/or NFκB in the cytoplasm for the proper control of cell survival if we were able to identify the exact mechanism by which these nuclear factors move.

Geldanamycin and geldanamycin-derivates are been currently used in clinical trials [175-177]. As discussed before, hsp90-inhibition affects the normal distribution of transcription factors, as well as their biological activity and stability in the cell [154,165,178]. In our hands, the immunophilin-binding drugs FK-506 and cyclosporine-A were ineffective for disrupting the hsp90•IMM-dependent mechanism of movement. However, it has recently been demonstrated that a ying-yang mechanism between FKBP51 and FKBP52 affects not only the steroid binding capacity of the GR but also its transcriptional activation [179-181]. Therefore, changes in subcellular distribution of proteins may be achieved by following this type of mechanism.

Dynamic nuclear mobility of proteins is essential for the normal progression of the cell cycle and the cell physiology itself. Accordingly, drugs or treatments that may affect this dynamic movement or the functional interactions among the components of the system may result revolutionary. The first step would be to elucidate whether there is a nuclear system for moving proteins to their sites of action, what the components are and how they are regulated. Unfortunately, our knowledge about how proteins move in the nucleus is just elemental and we are still far from solving these questions and, as is often the case, the answer may lie in the eye of the beholder, such that despite having very similar data, there are always notable differences in interpretation and emphasis. Though not necessarily incompatible, both the diffusion-dependent model and the targeted-based model for nuclear movement evoke long-standing questions as to the fundamental nature of the nucleoplasm itself. Our discussion of a still unsolved conundrum such as the molecular mechanism of movement of soluble proteins makes clear that many compelling questions remain and that explanations should not be expected to be simple. Rather, common ground may only be found in the complexity of the cytoplasmic and nuclear structure.

Acknowledgements

The experimental work described in this chapter was supported by grants from Consejo Nacional de Investigaciones Científicas y Técnicas de Argentina, Fundación Antorchas, and Agencia Nacional de Promoción Científica y Tecnológica (PICT 01-14123).

References

[1] Aridor, M. and Hannan, L.A. (2002) Traffic jams II: an update of diseases of intracellular transport. *Traffic* 3, 781-90.

[2] Cobbold, C., Monaco, A.P., Sivaprasadarao, A. and Ponnambalam, S. (2003) Aberrant trafficking of transmembrane proteins in human disease. *Trends Cell Biol* 13, 639-47.

[3] Gurevich, P., Zusman, I., Moldavsky, M., Szvalb, S., Elhayany, A., Halperin, R., Gurevich, E. and Ben-Hur, H. (2003) Secretory immune system in human intrauterine development: immunopathomorphological analysis of the role of secretory component (pIgR/SC) in immunoglobulin transport (review). *Int J Mol Med* 12, 289-97.

[4] Goldschneider, D., Blanc, E., Raguenez, G., Haddada, H., Benard, J. and Douc-Rasy, S. (2003) When p53 needs p73 to be functional - forced p73 expression induces nuclear accumulation of endogenous p53 protein. *Cancer Lett* 197, 99-103.

[5] Fabbro, M. and Henderson, B.R. (2003) Regulation of tumor suppressors by nuclear-cytoplasmic shuttling. *Exp Cell Res* 282, 59-69.

[6] Welch, W.J. (2004) Role of quality control pathways in human diseases involving protein misfolding. *Semin Cell Dev Biol* 15, 31-8.

[7] Evans, K.J., Gomes, E.R., Reisenweber, S.M., Gundersen, G.G. and Lauring, B.P. (2005) Linking axonal degeneration to microtubule remodeling by Spastin-mediated microtubule severing. *J Cell Biol* 168, 599-606.

[8] Cronshaw, J.M. and Matunis, M.J. (2004) The nuclear pore complex: disease associations and functional correlations. *Trends Endocrinol Metab* 15, 34-9.

[9] DeFranco, D.B. (2002) Navigating steroid hormone receptors through the nuclear compartment. *Mol Endocrinol* 16, 1449-55.

[10] Vicent, G.P., Pecci, A., Ghini, A., Piwien-Pilipuk, G. and Galigniana, M.D. (2002) Differences in nuclear retention characteristics of agonist-activated glucocorticoid receptor may determine specific responses. *Exp Cell Res* 276, 142-54.

[11] Xu, L. and Massague, J. (2004) Nucleocytoplasmic shuttling of signal transducers. *Nat Rev Mol Cell Biol* 5, 209-19.

[12] Fahrenkrog, B., Koser, J. and Aebi, U. (2004) The nuclear pore complex: a jack of all trades? *Trends Biochem Sci* 29, 175-82.

[13] Tang, J.X., Ito, T., Tao, T., Traub, P. and Janmey, P.A. (1997) Opposite effects of electrostatics and steric exclusion on bundle formation by F-actin and other filamentous polyelectrolytes. *Biochemistry* 36, 12600-7.

[14] Seksek, O., Biwersi, J. and Verkman, A.S. (1997) Translational diffusion of macromolecule-sized solutes in cytoplasm and nucleus. *J Cell Biol* 138, 131-42.

[15] Swaminathan, R., Bicknese, S., Periasamy, N. and Verkman, A.S. (1996) Cytoplasmic viscosity near the cell plasma membrane: translational diffusion of a small fluorescent solute measured by total internal reflection-fluorescence photobleaching recovery. *Biophys* J 71, 1140-51.

[16] Ellis, R.J. and Minton, A.P. (2003) Cell biology: join the crowd. *Nature* 425, 27-8.

[17] Goldstein, L.S. and Yang, Z. (2000) Microtubule-based transport systems in neurons: the roles of kinesins and dyneins. *Annu Rev Neurosci* 23, 39-71.

[18] Howe, C.L. and Mobley, W.C. (2004) Signaling endosome hypothesis: A cellular mechanism for long distance communication. *J Neurobiol* 58, 207-16.

[19] Ambron, R.T., Schmied, R., Huang, C.C. and Smedman, M. (1992) A signal sequence mediates the retrograde transport of proteins from the axon periphery to the cell body and then into the nucleus. *J Neurosci* 12, 2813-8.

[20] Galigniana, M.D., Radanyi, C., Renoir, J.M., Housley, P.R. and Pratt, W.B. (2001) Evidence that the peptidylprolyl isomerase domain of the hsp90-binding immunophilin FKBP52 is involved in both dynein interaction and glucocorticoid receptor movement to the nucleus. *J Biol Chem* 276, 14884-9.

[21] Piwien-Pilipuk, G. and Galigniana, M.D. (1998) Tautomycin inhibits phosphatase-dependent transformation of the rat kidney mineralocorticoid receptor. *Mol Cell Endocrinol* 144, 119-30.

[22] Thomas, M., Dadgar, N., Aphale, A., Harrell, J.M., Kunkel, R., Pratt, W.B. and Lieberman, A.P. (2004) Androgen receptor acetylation site mutations cause trafficking defects, misfolding, and aggregation similar to expanded glutamine tracts. *J Biol Chem* 279, 8389-95.

[23] Berg, P. and Pongratz, I. (2002) Two parallel pathways mediate cytoplasmic localization of the dioxin (aryl hydrocarbon) receptor. *J Biol Chem* 277, 32310-9.

[24] Barsony, J., Renyi, I. and McKoy, W. (1997) Subcellular distribution of normal and mutant vitamin D receptors in living cells. Studies with a novel fluorescent ligand. *J Biol Chem* 272, 5774-82.

[25] Lamb, J.R., Tugendreich, S. and Hieter, P. (1995) Tetratrico peptide repeat interactions: to TPR or not to TPR? *Trends Biochem Sci* 20, 257-9.

[26] Galat, A. (2003) Peptidylprolyl cis/trans isomerases (immunophilins): biological diversity--targets--functions. *Curr Top Med Chem* 3, 1315-47.

[27] Dahmer, M.K., Housley, P.R. and Pratt, W.B. (1984) Effects of molybdate and endogenous inhibitors on steroid-receptor inactivation, transformation, and translocation. *Annu Rev Physiol* 46, 67-81.

[28] Rousseau, G.G. (1984) Structure and regulation of the glucocorticoid hormone receptor. *Mol Cell Endocrinol* 38, 1-11.

[29] Pratt, W.B. (1987) Transformation of glucocorticoid and progesterone receptors to the DNA-binding state. *J Cell Biochem* 35, 51-68.

[30] Koyasu, S., Nishida, E., Kadowaki, T., Matsuzaki, F., Iida, K., Harada, F., Kasuga, M., Sakai, H. and Yahara, I. (1986) Two mammalian heat shock proteins, HSP90 and HSP100, are actin-binding proteins. *Proc Natl Acad Sci U S A* 83, 8054-8.

[31] Nishida, E., Koyasu, S., Sakai, H. and Yahara, I. (1986) Calmodulin-regulated binding of the 90-kDa heat shock protein to actin filaments. *J Biol Chem* 261, 16033-6.

[32] Miyata, Y. and Yahara, I. (1991) Cytoplasmic 8 S glucocorticoid receptor binds to actin filaments through the 90-kDa heat shock protein moiety. *J Biol Chem* 266, 8779-83.

[33] Gething, M.J. and Sambrook, J. (1992) Protein folding in the cell. Nature 355, 33-45.

[34] Wikstrom, A.C., Bakke, O., Okret, S., Bronnegard, M. and Gustafsson, J.A. (1987) Intracellular localization of the glucocorticoid receptor: evidence for cytoplasmic and nuclear localization. *Endocrinology* 120, 1232-42.

[35] Czar, M.J., Lyons, R.H., Welsh, M.J., Renoir, J.M. and Pratt, W.B. (1995) Evidence that the FK506-binding immunophilin heat shock protein 56 is required for trafficking of the glucocorticoid receptor from the cytoplasm to the nucleus. *Mol Endocrinol* 9, 1549-60.

[36] King, S.M. (2003) Organization and regulation of the dynein microtubule motor. *Cell Biol Int* 27, 213-5.

[37] Galigniana, M.D., Harrell, J.M., Murphy, P.J., Chinkers, M., Radanyi, C., Renoir, J.M., Zhang, M. and Pratt, W.B. (2002) Binding of hsp90-associated immunophilins to cytoplasmic dynein: direct binding and in vivo evidence that the peptidylprolyl isomerase domain is a dynein interaction domain. *Biochemistry* 41, 13602-10.

[38] Perrot-Applanat, M., Cibert, C., Geraud, G., Renoir, J.M. and Baulieu, E.E. (1995) The 59 kDa FK506-binding protein, a 90 kDa heat shock protein binding immunophilin (FKBP59-HBI), is associated with the nucleus, the cytoskeleton and mitotic apparatus. *J Cell Sci* 108 (Pt 5), 2037-51.

[39] Czar, M.J., Owens-Grillo, J.K., Yem, A.W., Leach, K.L., Deibel, M.R., Jr., Welsh, M.J. and Pratt, W.B. (1994) The hsp56 immunophilin component of untransformed steroid receptor complexes is localized both to microtubules in the cytoplasm and to the same nonrandom regions within the nucleus as the steroid receptor. *Mol Endocrinol* 8, 1731-41.

[40] Harrell, J.M., Murphy, P.J., Morishima, Y., Chen, H., Mansfield, J.F., Galigniana, M.D. and Pratt, W.B. (2004) Evidence for glucocorticoid receptor transport on microtubules by dynein. *J Biol Chem* 279, 54647-54.

[41] Galigniana, M.D., Harrell, J.M., O'Hagen, H.M., Ljungman, M. and Pratt, W.B. (2004) Hsp90-binding Immunophilins Link p53 to Dynein During p53 Transport to the Nucleus. *J Biol Chem* 279, 22483-22489.

[42] Reddy, R.K., Kurek, I., Silverstein, A.M., Chinkers, M., Breiman, A. and Krishna, P. (1998) High-molecular-weight FK506-binding proteins are components of heat-shock protein 90 heterocomplexes in wheat germ lysate. *Plant Physiol* 118, 1395-401.

[43] Harrell, J.M., Kurek, I., Breiman, A., Radanyi, C., Renoir, J.M., Pratt, W.B. and Galigniana, M.D. (2002) All of the protein interactions that link steroid receptor.hsp90.immunophilin heterocomplexes to cytoplasmic dynein are common to plant and animal cells. *Biochemistry* 41, 5581-7.

[44] Schena, M., Lloyd, A.M. and Davis, R.W. (1991) A steroid-inducible gene expression system for plant cells. *Proc Natl Acad Sci U S A* 88, 10421-5.

[45] Giannakakou, P., Sackett, D.L., Ward, Y., Webster, K.R., Blagosklonny, M.V. and Fojo, T. (2000) p53 is associated with cellular microtubules and is transported to the nucleus by dynein. *Nat Cell Biol* 2, 709-17.

[46] David-Pfeuty, T., Chakrani, F., Ory, K. and Nouvian-Dooghe, Y. (1996) Cell cycle-dependent regulation of nuclear p53 traffic occurs in one subclass of human tumor cells and in untransformed cells. *Cell Growth Differ* 7, 1211-25.

[47] Komarova, E.A., Zelnick, C.R., Chin, D., Zeremski, M., Gleiberman, A.S., Bacus, S.S. and Gudkov, A.V. (1997) Intracellular localization of p53 tumor suppressor protein in gamma-irradiated cells is cell cycle regulated and determined by the nucleus. *Cancer Res* 57, 5217-20.

[48] Bode, A.M. and Dong, Z. (2004) Post-translational modification of p53 in tumorigenesis. *Nat Rev Cancer* 4, 793-805.

[49] King, F.W., Wawrzynow, A., Hohfeld, J. and Zylicz, M. (2001) Co-chaperones Bag-1, Hop and Hsp40 regulate Hsc70 and Hsp90 interactions with wild-type or mutant p53. *Embo J* 20, 6297-305.

[50] Schliwa, M. and Woehlke, G. (2003) Molecular motors. *Nature* 422, 759-65.

[51] Hirokawa, N. (1998) Kinesin and dynein superfamily proteins and the mechanism of organelle transport. *Science* 279, 519-26.

[52] Burkhardt, J.K., Echeverri, C.J., Nilsson, T. and Vallee, R.B. (1997) Overexpression of the dynamitin (p50) subunit of the dynactin complex disrupts dynein-dependent maintenance of membrane organelle distribution. *J Cell Biol* 139, 469-84.

[53] Galigniana, M.D. and Piwien Pilipuk, G. (2004) Activation of the ligand-mineralocorticoid receptor functional unit by ancient, classical, and novel ligands. Structure-activity relationship. *Vitam Horm* 69, 31-68.

[54] Vicent, G.P., Pecci, A., Ghini, A.A., Piwien-Pilipuk, G., Veleiro, A.S., Burton, G., Lantos, C.P. and Galigniana, M.D. (1999) The glucocorticoid properties of the synthetic steroid pregna-1,4-diene-11beta-ol-3,20-dione (deltaHOP) are not entirely correlated with the steroid binding to the glucocorticoid receptor. *Mol Cell Endocrinol* 149, 207-19.

[55] Piwien-Pilipuk, G., Kanelakis, K.C., Ghini, A.A., Lantos, C.P., Litwack, G., Burton, G. and Galigniana, M.D. (2002) Modification of an essential amino group in the mineralocorticoid receptor evidences a differential conformational change of the receptor protein upon binding of antagonists, natural agonists and the synthetic agonist 11,19-oxidoprogesterone. *Biochim Biophys Acta* 1589, 31-48.

[56] Silverstein, A.M., Galigniana, M.D., Kanelakis, K.C., Radanyi, C., Renoir, J.M. and Pratt, W.B. (1999) Different regions of the immunophilin FKBP52 determine its association with the glucocorticoid receptor, hsp90, and cytoplasmic dynein. *J Biol Chem* 274, 36980-6.

[57] Denny, W.B., Valentine, D.L., Reynolds, P.D., Smith, D.F. and Scammell, J.G. (2000) Squirrel monkey immunophilin FKBP51 is a potent inhibitor of glucocorticoid receptor binding. *Endocrinology* 141, 4107-13.

[58] Scammell, J.G., Denny, W.B., Valentine, D.L. and Smith, D.F. (2001) Overexpression of the FK506-binding immunophilin FKBP51 is the common cause of glucocorticoid resistance in three New World primates. *Gen Comp Endocrinol* 124, 152-65.

[59] Antonin, W. and Mattaj, I.W. (2005) Nuclear pore complexes: round the bend? *Nat Cell Biol* 7, 10-2.

[60] Rabut, G., Lenart, P. and Ellenberg, J. (2004) Dynamics of nuclear pore complex organization through the cell cycle. *Curr Opin Cell Biol* 16, 314-21.

[61] Pante, N. (2004) Nuclear pore complex structure: unplugged and dynamic pores. *Dev Cell* 7, 780-1.

[62] Mosammaparast, N. and Pemberton, L.F. (2004) Karyopherins: from nuclear-transport mediators to nuclear-function regulators. *Trends Cell Biol* 14, 547-56.

[63] Stewart, M., Baker, R.P., Bayliss, R., Clayton, L., Grant, R.P., Littlewood, T. and Matsuura, Y. (2001) Molecular mechanism of translocation through nuclear pore complexes during nuclear protein import. *FEBS Lett* 498, 145-9.

[64] Bayliss, R., Littlewood, T. and Stewart, M. (2000) Structural basis for the interaction between FxFG nucleoporin repeats and importin-beta in nuclear trafficking. *Cell* 102, 99-108.

[65] Newmeyer, D.D. and Forbes, D.J. (1988) Nuclear import can be separated into distinct steps in vitro: nuclear pore binding and translocation. *Cell* 52, 641-53.

[66] Schwoebel, E.D., Ho, T.H. and Moore, M.S. (2002) The mechanism of inhibition of Ran-dependent nuclear transport by cellular ATP depletion. *J Cell Biol* 157, 963-74.

[67] Richardson, W.D., Mills, A.D., Dilworth, S.M., Laskey, R.A. and Dingwall, C. (1988) Nuclear protein migration involves two steps: rapid binding at the nuclear envelope followed by slower translocation through nuclear pores. *Cell* 52, 655-64.

[68] Jans, D.A., Xiao, C.Y. and Lam, M.H. (2000) Nuclear targeting signal recognition: a key control point in nuclear transport? *Bioessays* 22, 532-44.

[69] Harel, A. and Forbes, D.J. (2004) Importin beta: conducting a much larger cellular symphony. *Mol Cell* 16, 319-30.

[70] Ribbeck, K., Kutay, U., Paraskeva, E. and Gorlich, D. (1999) The translocation of transportin-cargo complexes through nuclear pores is independent of both Ran and energy. *Curr Biol* 9, 47-50.

[71] Kose, S., Imamoto, N., Tachibana, T., Shimamoto, T. and Yoneda, Y. (1997) Ran-unassisted nuclear migration of a 97-kD component of nuclear pore-targeting complex. *J Cell Biol* 139, 841-9.

[72] Englmeier, L., Olivo, J.C. and Mattaj, I.W. (1999) Receptor-mediated substrate translocation through the nuclear pore complex without nucleotide triphosphate hydrolysis. *Curr Biol* 9, 30-41.

[73] Xu, L., Kang, Y., Col, S. and Massague, J. (2002) Smad2 nucleocytoplasmic shuttling by nucleoporins CAN/Nup214 and Nup153 feeds TGFbeta signaling complexes in the cytoplasm and nucleus. *Mol Cell* 10, 271-82.

[74] Xu, L., Alarcon, C., Col, S. and Massague, J. (2003) Distinct domain utilization by Smad3 and Smad4 for nucleoporin interaction and nuclear import. *J Biol Chem* 278, 42569-77.

[75] Fagotto, F., Gluck, U. and Gumbiner, B.M. (1998) Nuclear localization signal-independent and importin/karyopherin-independent nuclear import of beta-catenin. *Curr Biol* 8, 181-90.

[76] Yokoya, F., Imamoto, N., Tachibana, T. and Yoneda, Y. (1999) beta-catenin can be transported into the nucleus in a Ran-unassisted manner. *Mol Biol Cell* 10, 1119-31.

[77] Matsubayashi, Y., Fukuda, M. and Nishida, E. (2001) Evidence for existence of a nuclear pore complex-mediated, cytosol-independent pathway of nuclear translocation of ERK MAP kinase in permeabilized cells. *J Biol Chem* 276, 41755-60.

[78] Whitehurst, A.W., Wilsbacher, J.L., You, Y., Luby-Phelps, K., Moore, M.S. and Cobb, M.H. (2002) ERK2 enters the nucleus by a carrier-independent mechanism. *Proc Natl Acad Sci U S A* 99, 7496-501.

[79] Hutchison, K.A., Czar, M.J. and Pratt, W.B. (1992) Evidence that the hormone-binding domain of the mouse glucocorticoid receptor directly represses DNA binding activity in a major portion of receptors that are "misfolded" after removal of hsp90. *J Biol Chem* 267, 3190-5.

[80] Hollenberg, S.M., Weinberger, C., Ong, E.S., Cerelli, G., Oro, A., Lebo, R., Thompson, E.B., Rosenfeld, M.G. and Evans, R.M. (1985) Primary structure and expression of a functional human glucocorticoid receptor cDNA. *Nature* 318, 635-41.

[81] Callebaut, I., Renoir, J.M., Lebeau, M.C., Massol, N., Burny, A., Baulieu, E.E. and Mornon, J.P. (1992) An immunophilin that binds M(r) 90,000 heat shock protein: main structural features of a mammalian p59 protein. *Proc Natl Acad Sci U S A* 89, 6270-4.

[82] Massol, N., Lebeau, M.C., Renoir, J.M., Faber, L.E. and Baulieu, E.E. (1992) Rabbit FKBP59-heat shock protein binding immunophillin (HBI) is a calmodulin binding protein. *Biochem Biophys Res Commun* 187, 1330-5.

[83] Jackson, D.A. and Cook, P.R. (1988) Visualization of a filamentous nucleoskeleton with a 23 nm axial repeat. *Embo J* 7, 3667-77.

[84] Davie, J.R., Samuel, S.K., Spencer, V.A., Holth, L.T., Chadee, D.N., Peltier, C.P., Sun, J.M., Chen, H.Y. and Wright, J.A. (1999) Organization of chromatin in cancer cells: role of signalling pathways. *Biochem Cell Biol* 77, 265-75.

[85] Carmo-Fonseca, M. and David-Ferreira, J.F. (1990) Interactions of intermediate filaments with cell structures. *Electron Microsc Rev* 3, 115-41.

[86] Czar, M.J., Welsh, M.J. and Pratt, W.B. (1996) Immunofluorescence localization of the 90-kDa heat-shock protein to cytoskeleton. *Eur J Cell Biol* 70, 322-30.

[87] Galigniana, M.D., Harrell, J.M., Housley, P.R., Patterson, C., Fisher, S.K. and Pratt, W.B. (2004) Retrograde transport of the glucocorticoid receptor in neurites requires dynamic assembly of complexes with the protein chaperone hsp90 and is linked to the CHIP component of the machinery for proteasomal degradation. *Brain Res Mol Brain Res* 123, 27-36.

[88] Roberts, B.L., Richardson, W.D. and Smith, A.E. (1987) The effect of protein context on nuclear location signal function. *Cell* 50, 465-75.

[89] Schneider, G. and Fechner, U. (2004) Advances in the prediction of protein targeting signals. *Proteomics* 4, 1571-80.

[90] Picard, D. and Yamamoto, K.R. (1987) Two signals mediate hormone-dependent nuclear localization of the glucocorticoid receptor. *Embo J* 6, 3333-40.

[91] Savory, J.G., Hsu, B., Laquian, I.R., Giffin, W., Reich, T., Hache, R.J. and Lefebvre, Y.A. (1999) Discrimination between NL1- and NL2-mediated nuclear localization of the glucocorticoid receptor. *Mol Cell Biol* 19, 1025-37.

[92] Urda, L.A., Yen, P.M., Simons, S.S., Jr. and Harmon, J.M. (1989) Region-specific antiglucocorticoid receptor antibodies selectively recognize the activated form of the ligand-occupied receptor and inhibit the binding of activated complexes to deoxyribonucleic acid. *Mol Endocrinol* 3, 251-60.

[93] Lei, E.P. and Silver, P.A. (2002) Protein and RNA export from the nucleus. *Dev Cel*l 2, 261-72.

[94] Berezney, R. (2002) Regulating the mammalian genome: the role of nuclear architecture. *Adv Enzyme Regul* 42, 39-52.

[95] Gilbert, N., Gilchrist, S. and Bickmore, W.A. (2005) Chromatin organization in the mammalian nucleus. *Int Rev Cytol* 242, 283-336.

[96] Zink, D., Fischer, A.H. and Nickerson, J.A. (2004) Nuclear structure in cancer cells. *Nat Rev Cancer* 4, 677-87.

[97] Gruenbaum, Y., Margalit, A., Goldman, R.D., Shumaker, D.K. and Wilson, K.L. (2005) The nuclear lamina comes of age. *Nat Rev Mol Cell Biol* 6, 21-31.

[98] Barrett, T.J. and Spelsberg, T.C. (1999) Nuclear matrix and steroid hormone action. *Vitam Horm* 55, 127-63.

[99] Bernhard, W. (1969) A new staining procedure for electron microscopical cytology. *J Ultrastruct Res* 27, 250-65.

[100] Monneron, A. and Bernhard, W. (1969) Fine structural organization of the interphase nucleus in some mammalian cells. *J Ultrastruct Res* 27, 266-88.

[101] Puvion, E. and Puvion-Dutilleul, F. (1996) Ultrastructure of the nucleus in relation to transcription and splicing: roles of perichromatin fibrils and interchromatin granules. *Exp Cell Res* 229, 217-25.

[102] Cmarko, D., Verschure, P.J., Martin, T.E., Dahmus, M.E., Krause, S., Fu, X.D., van Driel, R. and Fakan, S. (1999) Ultrastructural analysis of transcription and splicing in the cell nucleus after bromo-UTP microinjection. *Mol Biol Cell* 10, 211-23.

[103] Pederson, T. (2000) Half a century of "the nuclear matrix". *Mol Biol Cell* 11, 799-805.

[104] Mattern, K.A., Van Driel, R., and De Jong, L. (1996) Composition and structure of the internal nuclear matrix (In Monographies on Cell Biology: *Nuclear structure and gene* expression), pp. 87-110, Academic Press, New York.

[105] Tolstonog, G.V., Sabasch, M. and Traub, P. (2002) Cytoplasmic intermediate filaments are stably associated with nuclear matrices and potentially modulate their DNA-binding function. *DNA Cell Biol* 21, 213-39.

[106] Li, G., Tolstonog, G.V., Sabasch, M. and Traub, P. (2002) Interaction in vitro of type III intermediate filament proteins with supercoiled plasmid DNA and modulation of eukaryotic DNA topoisomerase I and II activities. *DNA Cell Biol* 21, 743-69.

[107] Tolstonog, G.V., Li, G., Shoeman, R.L. and Traub, P. (2005) Interaction in vitro of type III intermediate filament proteins with higher order structures of single-stranded DNA, particularly with G-quadruplex DNA. *DNA Cell Biol* 24, 85-110.

[108] Nickerson, J.A. (2001) Experimental observations of a nuclear matrix. *J. Cell Sci.* 114, 463-474.

[109] Htun, H., Barsony, J., Renyi, I., Gould, D.L. and Hager, G.L. (1996) Visualization of glucocorticoid receptor translocation and intranuclear organization in living cells with a green fluorescent protein chimera. *Proc Natl Acad Sci U S A* 93, 4845-50.

[110] Fejes-Toth, G., Pearce, D. and Naray-Fejes-Toth, A. (1998) Subcellular localization of mineralocorticoid receptors in living cells: effects of receptor agonists and antagonists. *Proc Natl Acad Sci U S A* 95, 2973-8.

[111] Tomura, A., Goto, K., Morinaga, H., Nomura, M., Okabe, T., Yanase, T., Takayanagi, R. and Nawata, H. (2001) The subnuclear three-dimensional image analysis of androgen receptor fused to green fluorescence protein. *J Biol Chem* 276, 28395-401.

[112] Prufer, K., Racz, A., Lin, G.C. and Barsony, J. (2000) Dimerization with retinoid X receptors promotes nuclear localization and subnuclear targeting of vitamin D receptors. *J Biol Chem* 275, 41114-23.

[113] Lim, C.S., Baumann, C.T., Htun, H., Xian, W., Irie, M., Smith, C.L. and Hager, G.L. (1999) Differential localization and activity of the A- and B-forms of the human progesterone receptor using green fluorescent protein chimeras. *Mol Endocrinol* 13, 366-75.

[114] Zhu, X.G., Hanover, J.A., Hager, G.L. and Cheng, S.Y. (1998) Hormone-induced translocation of thyroid hormone receptors in living cells visualized using a receptor green fluorescent protein chimera. *J Biol Chem* 273, 27058-63.

[115] Hart, L.L. and Davie, J.R. (2002) The estrogen receptor: more than the average transcription factor. *Biochem Cell Biol* 80, 335-41.

[116] Lauber, A.H., Sandhu, N.P., Schuchard, M., Subramaniam, M. and Spelsberg, T.C. (1995) Nuclear matrix acceptor binding sites for steroid hormone receptors: a candidate nuclear matrix acceptor protein. *Int Rev Cytol* 162B, 337-76.

[117] DeFranco, D.B. and Guerrero, J. (2000) Nuclear matrix targeting of steroid receptors: specific signal sequences and acceptor proteins. *Crit Rev Eukaryot Gene Expr* 10, 39-44.

[118] Beato, M. (1996) Chromatin structure and the regulation of gene expression: remodeling at the MMTV promoter. *J Mol Med* 74, 711-24.

[119] Beato, M., Chavez, S. and Truss, M. (1996) Transcriptional regulation by steroid hormones. *Steroids* 61, 240-51.

[120] Georget, V., Terouanne, B., Nicolas, J.C. and Sultan, C. (2002) Mechanism of antiandrogen action: key role of hsp90 in conformational change and transcriptional activity of the androgen receptor. *Biochemistry* 41, 11824-31.

[121] Hendzel, M.J., Kruhlak, M.J., MacLean, N.A., Boisvert, F., Lever, M.A. and Bazett-Jones, D.P. (2001) Compartmentalization of regulatory proteins in the cell nucleus. *J Steroid Biochem Mol Biol* 76, 9-21.

[122] Hirst, M.A., Northrop, J.P., Danielsen, M. and Ringold, G.M. (1990) High level expression of wild type and variant mouse glucocorticoid receptors in Chinese hamster ovary cells. *Mol Endocrinol* 4, 162-70.

[123] Tago, K., Tsukahara, F., Naruse, M., Yoshioka, T. and Takano, K. (2004) Regulation of nuclear retention of glucocorticoid receptor by nuclear Hsp90. *Mol Cell Endocrinol* 213, 131-8.

[124] Elbi, C., Walker, D.A., Romero, G., Sullivan, W.P., Toft, D.O., Hager, G.L. and DeFranco, D.B. (2004) Molecular chaperones function as steroid receptor nuclear mobility factors. *Proc Natl Acad Sci U S A* 101, 2876-81.

[125] Freeman, B.C., Felts, S.J., Toft, D.O. and Yamamoto, K.R. (2000) The p23 molecular chaperones act at a late step in intracellular receptor action to differentially affect ligand efficacies. *Genes Dev* 14, 422-34.

[126] Freeman, B.C. and Yamamoto, K.R. (2002) Disassembly of transcriptional regulatory complexes by molecular chaperones. *Science* 296, 2232-5.

[127] Tang, Y. and DeFranco, D.B. (1996) ATP-dependent release of glucocorticoid receptors from the nuclear matrix. *Mol Cell Biol* 16, 1989-2001.

[128] Oxelmark, E., Knoblauch, R., Arnal, S., Su, L.F., Schapira, M. and Garabedian, M.J. (2003) Genetic dissection of p23, an Hsp90 cochaperone, reveals a distinct surface involved in estrogen receptor signaling. *J Biol Chem* 278, 36547-55.

[129] Morishima, Y., Kanelakis, K.C., Murphy, P.J., Lowe, E.R., Jenkins, G.J., Osawa, Y., Sunahara, R.K. and Pratt, W.B. (2003) The hsp90 cochaperone p23 is the limiting component of the multiprotein hsp90/hsp70-based chaperone system in vivo where it acts to stabilize the client protein: hsp90 complex. *J Biol Chem* 278, 48754-63.

[130] Felts, S.J. and Toft, D.O. (2003) p23, a simple protein with complex activities. *Cell Stress Chaperones* 8, 108-13.

[131] Sullivan, W.P., Owen, B.A. and Toft, D.O. (2002) The influence of ATP and p23 on the conformation of hsp90. *J Biol Chem* 277, 45942-8.

[132] Hager, G.L., Elbi, C. and Becker, M. (2002) Protein dynamics in the nuclear compartment. *Curr Opin Genet Dev* 12, 137-41.

[133] Elbi, C., Walker, D.A., Lewis, M., Romero, G., Sullivan, W.P., Toft, D.O., Hager, G.L. and DeFranco, D.B. (2004) A novel in situ assay for the identification and characterization of soluble nuclear mobility factors. *Sci STKE* 2004, pl10.

[134] Hendzel, M.J., Boisvert, F. and Bazett-Jones, D.P. (1999) Direct visualization of a protein nuclear architecture. *Mol Biol Cell* 10, 2051-62.

[135] Richter, K., Reichenzeller, M., Gorisch, S.M., Schmidt, U., Scheuermann, M.O., Herrmann, H. and Lichter, P. (2005) Characterization of a nuclear compartment shared by nuclear bodies applying ectopic protein expression and correlative light and electron microscopy. *Exp Cell Res* 303, 128-37.

[136] Markova, M.D. (2004) Electron microscopic observations of human sperm whole-mounts after extraction for nuclear matrix and intermediate filaments (NM-IF). *Int J Androl* 27, 291-5.

[137] Miccoli, L., Biard, D.S., Frouin, I., Harper, F., Maga, G. and Angulo, J.F. (2003) Selective interactions of human kin17 and RPA proteins with chromatin and the nuclear matrix in a DNA damage- and cell cycle-regulated manner. *Nucleic Acids Res* 31, 4162-75.

[138] Gajkowska, B. and Wojewodzka, U. (2003) A new look at the cellular scaffold by embedment-free electron microscopy method. *J Cell Mol Med* 7, 258-64.

[139] Li, G., Tolstonog, G.V., Sabasch, M. and Traub, P. (2003) Type III intermediate filament proteins interact with four-way junction DNA and facilitate its cleavage by the junction-resolving enzyme T7 endonuclease I. *DNA Cell Biol* 22, 261-91.

[140] Mintz, P.J., Patterson, S.D., Neuwald, A.F., Spahr, C.S. and Spector, D.L. (1999) Purification and biochemical characterization of interchromatin granule clusters. *Embo J* 18, 4308-20.

[141] Nickerson, J.A., Krockmalnic, G., Wan, K.M. and Penman, S. (1997) The nuclear matrix revealed by eluting chromatin from a cross-linked nucleus. *Proc Natl Acad Sci USA* 94, 4446-50.

[142] Wan, K.M., Nickerson, J.A., Krockmalnic, G. and Penman, S. (1999) The nuclear matrix prepared by amine modification. *Proc Natl Acad Sci U S A* 96, 933-8.

[143] Engelhardt, M. (1999) Demonstration of a DNase-sensitive network associated with the nuclear pore complexes in rat liver nuclei. *Chromosoma* 108, 64-71.

[144] Glazko, G.V., Koonin, E.V., Rogozin, I.B. and Shabalina, S.A. (2003) A significant fraction of conserved noncoding DNA in human and mouse consists of predicted matrix attachment regions. *Trends Genet* 19, 119-24.

[145] Bode, J., Goetze, S., Heng, H., Krawetz, S.A. and Benham, C. (2003) From DNA structure to gene expression: mediators of nuclear compartmentalization and dynamics. *Chromosome Res* 11, 435-45.

[146] Scott, R.E., Giannakouros, T., Gao, S. and Peidis, P. (2003) Functional potential of P2P-R: a role in the cell cycle and cell differentiation related to its interactions with proteins that bind to matrix associated regions of DNA? *J Cell Biochem* 90, 6-12.

[147] Spector, D.L. (1996) Nuclear organization and gene expression. *Exp Cell Res* 229, 189-97.

[148] Zeng, C., van Wijnen, A.J., Stein, J.L., Meyers, S., Sun, W., Shopland, L., Lawrence, J.B., Penman, S., Lian, J.B., Stein, G.S. and Hiebert, S.W. (1997) Identification of a nuclear matrix targeting signal in the leukemia and bone-related AML/CBF-alpha transcription factors. *Proc Natl Acad Sci U S A* 94, 6746-51.

[149] Stenoien, D.L., Patel, K., Mancini, M.G., Dutertre, M., Smith, C.L., O'Malley, B.W. and Mancini, M.A. (2001) FRAP reveals that mobility of oestrogen receptor-alpha is ligand- and proteasome-dependent. *Nat Cell Biol* 3, 15-23.

[150] Gorisch, S.M., Richter, K., Scheuermann, M.O., Herrmann, H. and Lichter, P. (2003) Diffusion-limited compartmentalization of mammalian cell nuclei assessed by microinjected macromolecules. *Exp Cell Res* 289, 282-94.

[151] Misteli, T., Caceres, J.F. and Spector, D.L. (1997) The dynamics of a pre-mRNA splicing factor in living cells. *Nature* 387, 523-7.

[152] Kruhlak, M.J., Lever, M.A., Fischle, W., Verdin, E., Bazett-Jones, D.P. and Hendzel, M.J. (2000) Reduced mobility of the alternate splicing factor (ASF) through the nucleoplasm and steady state speckle compartments. *J Cell Biol* 150, 41-51.

[153] Meier, U.T. and Blobel, G. (1992) Nopp140 shuttles on tracks between nucleolus and cytoplasm. *Cell* 70, 127-38.

[154] Pratt, W.B. and Toft, D.O. (1997) Steroid receptor interactions with heat shock protein and immunophilin chaperones. *Endocr Rev* 18, 306-60.

[155] Parada, L. and Misteli, T. (2002) Chromosome positioning in the interphase nucleus. *Trends Cell Biol* 12, 425-32.

[156] Cremer, T. and Cremer, C. (2001) Chromosome territories, nuclear architecture and gene regulation in mammalian cells. *Nat Rev Genet* 2, 292-301.

[157] Jackson, D.A. (2003) The principles of nuclear structure. *Chromosome Res* 11, 387-401.

[158] Cremer, T., Kurz, A., Zirbel, R., Dietzel, S., Rinke, B., Schrock, E., Speicher, M.R., Mathieu, U., Jauch, A., Emmerich, P. and et al. (1993) Role of chromosome territories in the functional compartmentalization of the cell nucleus. *Cold Spring Harb Symp Quant Biol* 58, 777-92.

[159] Verschure, P.J., van Der Kraan, I., Manders, E.M. and van Driel, R. (1999) Spatial relationship between transcription sites and chromosome territories. *J Cell Biol* 147, 13-24.

[160] Volpi, E.V., Chevret, E., Jones, T., Vatcheva, R., Williamson, J., Beck, S., Campbell, R.D., Goldsworthy, M., Powis, S.H., Ragoussis, J., Trowsdale, J. and Sheer, D. (2000) Large-scale chromatin organization of the major histocompatibility complex and other regions of human chromosome 6 and its response to interferon in interphase nuclei. *J Cell Sci* 113 (Pt 9), 1565-76.

[161] Piwien Pilipuk, G., Galigniana, M.D. and Schwartz, J. (2003) Subnuclear localization of C/EBP beta is regulated by growth hormone and dependent on MAPK. *J Biol Chem* 278, 35668-77.

[162] Liao, J., Piwien-Pilipuk, G., Ross, S.E., Hodge, C.L., Sealy, L., MacDougald, O.A. and Schwartz, J. (1999) CCAAT/enhancer-binding protein beta (C/EBPbeta) and C/EBPdelta contribute to growth hormone-regulated transcription of c-fos. *J Biol Chem* 274, 31597-604.

[163] Piwien-Pilipuk, G. (2003) Diverse functions of CCAAT/enhancer binding proteins are achieved by different regulatory mechanisms. *Curr Topics in Biochem Res* 5, 161-170.

[164] Schaaf, M.J. and Cidlowski, J.A. (2003) Molecular determinants of glucocorticoid receptor mobility in living cells: the importance of ligand affinity. *Mol Cell Biol* 23, 1922-34.

[165] Stavreva, D.A., Muller, W.G., Hager, G.L., Smith, C.L. and McNally, J.G. (2004) Rapid glucocorticoid receptor exchange at a promoter is coupled to transcription and regulated by chaperones and proteasomes. *Mol Cell Biol* 24, 2682-97.

[166] Galigniana, M.D., Morishima, Y., Gallay, P.A. and Pratt, W.B. (2004) Cyclophilin-A is bound through its peptidylprolyl isomerase domain to the cytoplasmic dynein motor protein complex. *J Biol Chem* 279, 55754-9.

[167] McDonald, D., Vodicka, M.A., Lucero, G., Svitkina, T.M., Borisy, G.G., Emerman, M. and Hope, T.J. (2002) Visualization of the intracellular behavior of HIV in living cells. *J Cell Biol* 159, 441-52.

[168] Minder, D., Boni, J., Schupbach, J. and Gehring, H. (2002) Immunophilins and HIV-1 infection. *Arch Virol* 147, 1531-42.

[169] Wiegers, K. and Krausslich, H.G. (2002) Differential dependence of the infectivity of HIV-1 group O isolates on the cellular protein cyclophilin A. *Virology* 294, 289-95.

[170] Dietrich, L., Ehrlich, L.S., LaGrassa, T.J., Ebbets-Reed, D. and Carter, C. (2001) Structural consequences of cyclophilin A binding on maturational refolding in human immunodeficiency virus type 1 capsid protein. *J Virol* 75, 4721-33.

[171] Samuel, S.K., Spencer, V.A., Bajno, L., Sun, J.M., Holth, L.T., Oesterreich, S. and Davie, J.R. (1998) In situ cross-linking by cisplatin of nuclear matrix-bound transcription factors to nuclear DNA of human breast cancer cells. *Cancer Res* 58, 3004-8.

[172] Spencer, V.A., Samuel, S.K. and Davie, J.R. (2000) Nuclear matrix proteins associated with DNA in situ in hormone-dependent and hormone-independent human breast cancer cell lines. *Cancer Res* 60, 288-92.

[173] Karin, M., Cao, Y., Greten, F.R. and Li, Z.W. (2002) NF-kappaB in cancer: from innocent bystander to major culprit. *Nat Rev Cancer* 2, 301-10.

[174] Chene, P. (2003) Inhibiting the p53-MDM2 interaction: an important target for cancer therapy. *Nat Rev Cancer* 3, 102-9.

[175] Neckers, L. (2002) Hsp90 inhibitors as novel cancer chemotherapeutic agents. *Trends Mol Med* 8, S55-61.

[176] Workman, P. (2004) Altered states: selectively drugging the Hsp90 cancer chaperone. *Trends Mol Med* 10, 47-51.

[177] Uehara, Y. (2003) Natural product origins of Hsp90 inhibitors. *Curr Cancer Drug Targets* 3, 325-30.

[178] Richter, K. and Buchner, J. (2001) Hsp90: chaperoning signal transduction. *J Cell Physiol* 188, 281-90.

[179] Wochnik, G.M., Ruegg, J., Abel, G.A., Schmidt, U., Holsboer, F. and Rein, T. (2005) FK506-binding proteins 51 and 52 differentially regulate dynein interaction and nuclear translocation of the glucocorticoid receptor in mammalian cells. *J Biol Chem* 280, 4609-16.

[180] Hubler, T.R., Denny, W.B., Valentine, D.L., Cheung-Flynn, J., Smith, D.F. and Scammell, J.G. (2003) The FK506-binding immunophilin FKBP51 is transcriptionally regulated by progestin and attenuates progestin responsiveness. *Endocrinology* 144, 2380-7.

[181] Davies, T.H., Ning, Y.M. and Sanchez, E.R. (2002) A new first step in activation of steroid receptors: hormone-induced switching of FKBP51 and FKBP52 immunophilins. *J Biol Chem* 277, 4597-600.

In: Focus on Cellular Signalling Research
Editor: Dorothy T. Leeds, pp. 109-129
ISBN 1-59454-619-3

Chapter V

Two Is Better than One: Structure, Function and Mechanism of Tandem Domains

Amjad Farooq[1,2,3*], ***Marius Sudol***[4,5] ***and Ming-Ming Zhou***[6]
[1]Department of Biochemistry and Molecular Biology, University of Miami Miller School of Medicine, 1011 NW 15th Street, Miami, FL 33136
[2]Braman Family Breast Cancer Institute, Sylvester Comprehensive Cancer Center, University of Miami Miller School of Medicine, Miami, FL 33136
[4]Weis Center for Research, Geisinger Clinic, 100 North Academy Avenue, Danville, PA 17822
[5]Department of Medicine, Mount Sinai School of Medicine, One Gustave L.Levy Place, New York, NY 10029
[6]Structural Biology Program, Department of Physiology and Biophysics, Mount Sinai School of Medicine, One Gustave L.Levy Place, New York, NY 10029

Abstract

The specificity and fidelity in cellular signal transduction is in part owed to the remarkable ability of signaling and adaptor proteins to exist in modular fashion. Such modular organization of proteins allows them to participate in a wide spectrum of signaling circuits simultaneously yet in a highly specific manner. This is the result of the specific interaction between protein modules found in a wide multitude of proteins and tailored to meet specific demands. Recent studies indicate that such protein modules may not always act independently in a one-on-one manner but rather their activity may be enhanced by the presence of a second copy in tandem. Such tandem repeats appear to act in a highly synergistic manner so that their biological function is greater than the sum of their parts. Using bromodomains, SH2 domains, WW domains and PDZ domains as prototypes, we discuss the structure, function and mechanism of how such tandem

* Corresponding author: amjad@farooqlab.ws; tel 305-243-2429; fax 305-243-3955.

repeats act in a co-operative manner and the biological implications of such an elegant design that nature seems to have adopted.

Introduction

Proteins involved in the regulation of cellular processes such as cell growth and differentiation, embryogenic development, cell cycle regulation and apoptosis are generally constructed in a modular fashion from a combination of regulatory and catalytic domains. Our current understanding of such a modular organization of proteins is deeply indebted to the pioneering work of Pawson and co-workers during the 1980s. In a landmark study, they demonstrated that a region of about 100 residues located at the N-terminus of the viral tyrosine kinase v-Fps/Fes was not required for its catalytic activity yet its truncation abolished its oncogenic potential [1]. This element, that has come to be known as the SH2 domain, is present in a wide variety of cellular proteins and its discovery laid the foundations for the modular organization of proteins — a feature that is the hallmark of cellular signal transduction. To date, the number of distinct protein modules identified across a diverse spectrum of cellular proteins inarguably runs in three digits, underlying the remarkable explosion of information on the modular behavior of proteins over the last two decades. Some of the best studied examples of protein modules include the inaugural SH2 domain, the PTB domain, the SH3 domain, the WW domain, the PDZ domain and the bromodomain.

The construction of proteins in a modular fashion is a remarkable virtue that enables proteins not only to act as adaptors and link multiple partners together but also allows them to engage in cross-talk across a diverse array of signaling circuits [2, 3]. The localization of proteins at specific locations within the cell coupled with the transient nature of interactions between specific protein modules further ensures that the temporal and spatial specificity of signaling machinery is not compromised [4, 5]. Since the discovery of the SH2 domain in the mid 1980s [1], structural and functional analysis of a wide range of protein modules has significantly added to our understanding of their mechanism of action at molecular level. More recently, our continuing efforts to gain a deeper insight into the mechanisms of cellular signal transduction have led to a surprising discovery: Although capable of acting independently, protein modules may not be all that independent. In many proteins, multiple copies of the same domain, often arranged in tandem, are also found. Several lines of evidence suggest that such tandemly arranged protein domains do not act independently but rather modulate the activity of each other in a cooperative manner — the tandem repeats bind to their ligands with an affinity that may be several orders of magnitude greater than to individual domains. For example, the tandem bromodomains of the transcription factor TAFII250 bind acetylated peptides with an affinity nearly 100-fold higher than that observed for individual bromodomains [6]. In the case of SH2 domains of ZAP70 tyrosine kinase [7], Syk tyrosine kinase [8], phosphatidylinositol-3-kinase [9], SHP2 tyrosine phosphatase [10, 11] and phospholipase Cγ1 [12], the binding affinity of phosphorylated peptides to tandem SH2 domains is over 1000-fold greater relative to isolated SH2 domains. The phenomenon of the modulation of the activity of one domain by another in a tandem repeat is also exquisitely demonstrated by the PDZ domains of several proteins including PSD95 [13], syntenin [14, 15], X11 [16] and GRIP [17, 18].

Protein modules have been known to modulate the activity of their catalytic domains for some time. For example, the SH2 domain blocks the active site of the catalytic domain of SHP2 phosphatase and only upon the interaction of the SH2 domain with phosphorylated receptor tails is this interaction relieved and thereby allowing the catalytic domain to retain its full activity [10]. In a similar manner, the PH domain of the Akt/PKB kinase also blocks the kinase active site [19]. Recruitment of Akt/PKB kinase to the membrane surface via interaction between its PH domain and membrane phospholipids subsequently relieves the inhibitory action of the PH domain. The demonstration that protein modules modulate each other's activity, especially in tandem repeats, is however a relatively novel phenomenon and has caught our attention. In this review, we discuss the molecular basis of the modulation of the activity of one protein module by another in a tandem repeat using bromodomains, SH2 domains, WW domains and PDZ domains as prototypes.

Tandem Bromodomains

Although DNA contains the genetic information necessary for the synthesis of proteins, whether this information is actually used at any given time depends on the state of histones that organize the DNA into an higher order assembly known as chromatin. Histones are subject to a dazzling array of site-specific post-translational modifications such as acetylation, methylation, phosphorylation, ubiquitination, sumoylation and ADP-ribosylation [20, 21]. Several lines of evidence indicate that the state of histone modification is an important determinant in switching gene expression on or off — a phenomenon that has come to be known as histone code [22, 23]. Histone modification induces structural changes within the chromatin and thus influences whether a gene can be transcribed or not. Although eukaryotic cells contain an arsenal of enzymes that can add or remove specific modifications in histones, additional protein modules have also been identified that indirectly increase the efficiency of such enzymatically-driven modifications in chromatin. One such protein module is the bromodomain — a module found in many chromatin-associated proteins and nearly all nuclear histone acetyltransferases (HATs). Upon histone acetylation by HATs, bromodomains are believed to bind the acetylated lysines within histones and thereby lead to highly specific acetylation of histones by tethering HATs to specific chromosomal sites [24, 25]. Bromodomains thus represent an important part of the transcriptional machinery involved in chromatin remodeling and gene expression (Figure 1a).

Although bromodomains were originally identified as autonomous protein modules in PCAF [25] and GCN5 [26, 27], their occurrence as tandem repeats is also well documented in a variety of proteins from humans and other eukaryotes (Figure 1b). Examples include the human TAFII250 [6], the mammalian Brd2 and Brd4 [28, 29], the fly Fsh [30], and the yeast Bdf1 and Bdf2 [31, 32]. It is now well understood that the tandem bromodomains act in a co-operative manner that results in binding to acetylated peptides, derived from their potential binding sites in histones, with an affinity that is nearly 100-fold greater than that observed for individual bromodomains [6].

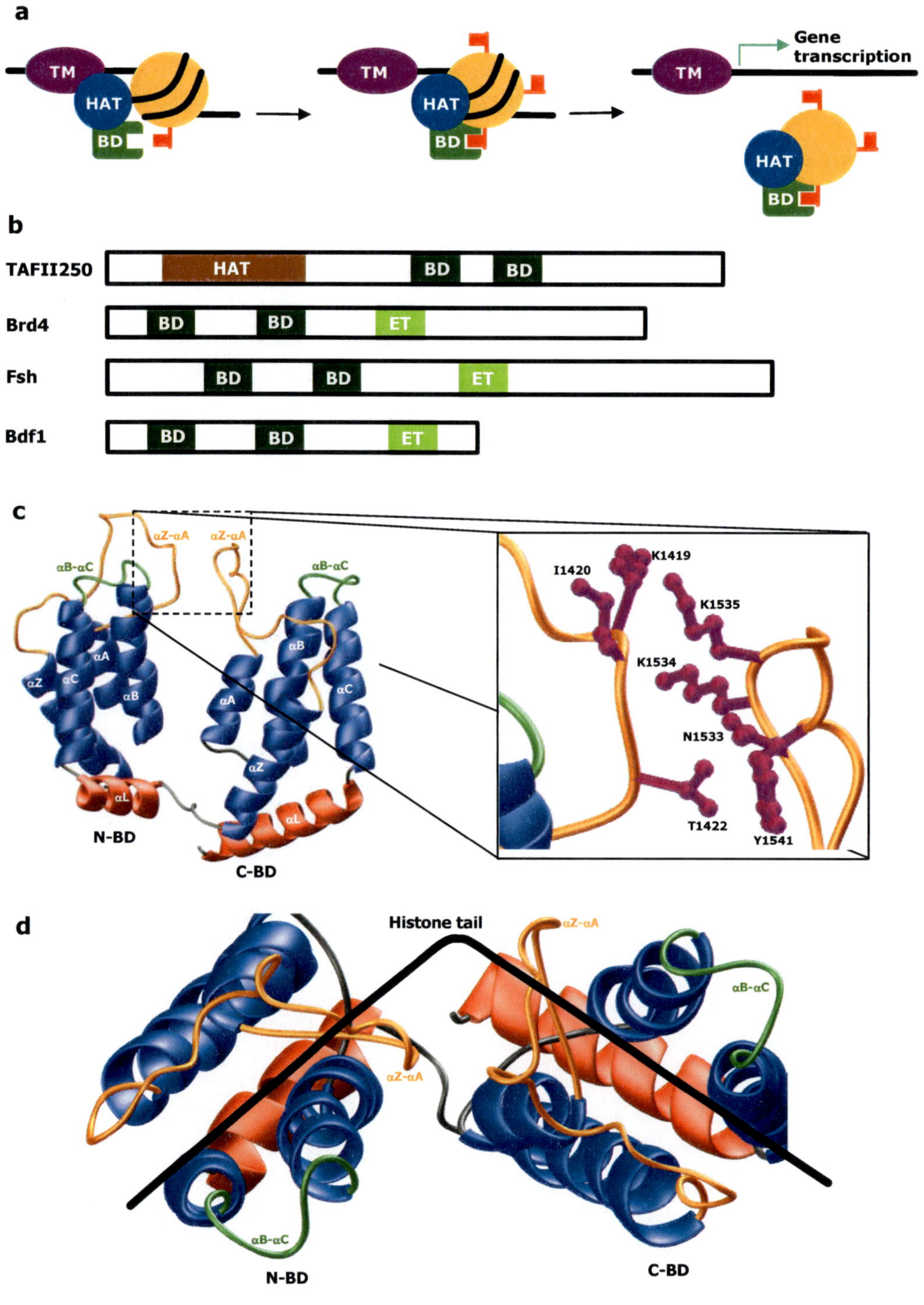

Figure 1. Tandem bromodomains. (a) A schematic showing the involvement of bromodomain (BD) in tethering histone acetyltransferase (HAT) to chromatin. In this scheme, the binding of transcriptional machinery (TM) to the TATA box within the promoter would recruit HAT. Histone acetylation by HAT in turn would allow the BD to bind to acetylated histones and thereby leading to further acetylation of histones resulting in re-organization of chromatin so as to allow the TM to transcribe the

part of DNA unavailable prior to histone acetylation. (b) Examples of some proteins containing tandem repeats of bromodomain (BD). (c) Crystal structure of the tandem bromodomain of the TAFII250 subunit of the transcription factor TAFIID. The N-terminal and the C-terminal domains are respectively labeled N-BD and C-BD. Helices αZ, αA, αB and αC that constitute a canonical bromodomain are colored blue, the helices αL that are found at the C-termini of each domain are colored red and, the loops αZ-αA and αB-αC that constitute the peptide binding pocket are respectively colored yellow and green. The inset shows the interdomain interactions between the sidechains of key residues located within the loops αZ-αA. (d) Top view of the tandem bromodomain shown in (c) providing a closer look at the peptide binding sites constituted by the loops αZ-αA and αB-αC. Hand-drawn hypothetical orientation and positioning of H4 peptide is indicated by a thick black L-shaped line.

To understand the molecular basis of such an heightened affinity and specificity, Tjian and co-workers solved the 3D structure of the tandem bromodomain of the TAFII250 subunit of the transcription factor TAFIID [6]. Each bromodomain within the U-shaped tandem repeat is comprised of a four-helical bundle with a left-handed twist (Figure 1c). This topology was originally observed in the structure of PCAF bromodomain [25]. The four helices, labeled respectively αZ, αA, αB and αC from the N-terminus to the C-terminus, lie roughly anti-parallel to each other and the loops αZ-αA and αB-αC are positioned within each other's vicinity at one end of the four-helical bundle. The N-terminal bromodomain (N-BD) is linked to the C-terminal bromodomain (C-BD) through an additional helix αL that lies C-terminal to the helix αC of N-BD and N-terminal to helix αZ of C-BD. This arrangement gives the two bromodomains an overall U-shaped architecture. The helix αL is also present at the C-terminus of helix αC of C-BD but its role is not clear. The N-BD and the C-BD superimpose upon each other with a Cα RMSD of ~ 0.8 Å, implying that the two bromodomains essentially adopt an identical 3D structure. The N-BD and the C-BD domains are held together at two distinct contact points. These include electrostatic interactions between E1443, E1444 and E1447 in helix αB of N-BD and K1555 and K1559 in helix αA of C-BD. In addition, hydrophobic interactions between residues lining the loops αZ-αA within each domain ensure that the two domains are locked into a fixed orienation relative to each other (Figure 1c, inset). The two putative binding pockets for acetylated lysines are largely formed by hydrophobic residues located within the loops αZ-αA and αB-αC within each bromodomain (Figure 1d). The U-shaped architecture of the tandem bromodomains places these binding pockets at an optimal distance and orientation to favor the binding of a bidentate ligand in a synergistic manner. The distance between the two binding pockets of ~25 Å suggests that the tandem bromodomains are ideally suited to bind acetylated lysines separated by about 7-8 residues — a feature that is frequently encountered within histone tails. In addition, the two binding pockets lie roughly perpendicular to each other, implying that the bidentate ligand would assume an L-shaped conformation upon binding to the tandem bromodomains.

In summary, the arrangement of tandem bromodomains of TAFII250 in a U-shaped structure not only helps them to optimally interact with multiply acetylated histone tails but also increases their affinity for their targets by nearly two orders of magnitude [6]. Such a feature would clearly render tandem bromodomains more efficient at causing structural changes within chromatin that ultimately determine the outcome of gene expression. Identification of specific substrates of TAFII250 bromodomains would clearly shed further light on the precise mechanism of action of the tandem bromodomains of TAFII250.

Tandem SH2 Domains

Cell surface receptors, such as receptor tyrosine kinases (RTKs), are the cell's equivalent of antennae that transmit extracellular information in the form of growth factors, antigens, cytokines and so on at the plasma membrane to specific distant cellular targets such as transcription factors within the nucleus [2, 3]. Such chemical messengers activate RTKs, enabling them to bind a plethora of protein modules within the cytosol with distinct specificities and biological functions [33]. One such module is the SH2 domain, which upon binding to the cytoplasmic tail of the activated RTKs relays the signal to specific downstream targets involved in cellular activities such as cell growth, cell differentiation and apoptosis. SH2 domains specifically recognize phosphotyrosine (pY) and residues that lie C-terminal to pY within the cytoplasmic tails of RTKs as well as a wide range of other adaptor and docking proteins within the cytosol. For example, binding of Grb2 SH2 domain to the pYXN motif within activated receptors, such as EGFR [34], enables the Grb2 adaptor protein to interact with the guanine nucleotide exchange factor SOS, through its N-terminal SH3 domain, leading to Ras activation — a protein that orchestrates and links many different signaling pathways [35, 36] (Figure 2a). Although SH2 domains are widely found as single copies in adaptor and docking proteins such as Grb2 and Shc [34, 36], it is their occurrence as tandem repeats in a wide variety of enzymes that is really intriguing. These include ZAP70 tyrosine kinase [7], Syk tyrosine kinase [8], phosphatidylinositol-3-kinase [9], SHP2 tyrosine phosphatase [11] and phospholipase Cγ1 [12] (Figure 2b). The tandem SH2 domains within these proteins do not act independently but rather modulate the activity of each other through synergistic effect — the binding affinity of tyrosine-phosphorylated peptides is between 100-1000 fold greater to tandem SH2 domains relative to individual SH2 domains. In an attempt to unravel the molecular mechanism of how cognate peptides confer greater affinity upon tandem SH2 domains, structures of SH2 domains of ZAP70 tyrosine kinase [7], Syk tyrosine kinase [8] and SHP2 tyrosine phosphatase [11] complexed to their bidentate peptide ligands have been solved. In order to maintain brevity, we discuss the structure of tandem SH2 domains of Syk tyrosine kinase in complex with a bidentate tyrosine-phosphorylated peptide [8] to illustrate the general principles of high affinity interaction of tandem SH2 domains with their cognate ligands.

Syk tyrosine kinase specifically phosphorylates tyrosine residues located within the bidentate consensus motif YXXZ-X_n-YXXZ frequently found in signaling components of antigen and antibody receptors [37]. Within this consensus sequence, X is any residue, Z is usually an hydrophobic residue such as leucine or isoleucine, and n is typically 7-8 residues. Phosphorylation of tyrosine within these motifs allows the binding of the tandem SH2 domains leading to the recruitment of Syk tyrosine kinase to the sites of antigen and antibody activity [38]. Each SH2 domain within the tandem repeat of Syk tyrosine kinase is comprised of a central anti-parallel β-sheet (strands βA-βG) sandwiched between two helices αA and αB on each face (Figure 2c). The central β-sheet resembles the shape of a baseball glove with the helix αB more or less located at a position inside the glove where the baseball would sit and the helix αA sitting on the back of the glove. This topology is characteristic of the evolutionarily conserved SH2 fold first observed in the crystal structure of SH2 domain of v-Src tyrosine kinase determined by Kuriyan and co-workers [39]. The N-terminal (N-SH2) and the C-terminal (C-SH2) canonical domains in Syk tyrosine kinase are glued together by a stem comprised of three helices αC, αD and αE, that originate from the C-terminal side of the last strand βG of N-SH2 and terminate at N-terminal of the first strand βA of C-SH2, giving the tandem SH2 domains an overall Y-shaped architecture. Such an architecture is also observed in the tandem SH2 domains of ZAP70 tyrosine kinase [7] but not in SHP2 tyrosine phosphatase [11]. In the latter, the SH2 tandem domains are held together by a short interdomain linker sequence that gives the molecule an overall V-shaped look.

In addition to the orientational constraints imposed by the three-helical stem, the SH2 tandem domains of Syk tyrosine kinase are further co-stabilized through a complex network of interdomain interactions between residues located at the N-terminus of helix αA and the loops βA-αA and βB-βC of the N-SH2 domain, and helix αB, strand βF and the loops βF-αB and αB-βG of the C-SH2 domain (Figure 2c, inset).

The tyrosine-phosphorylated peptide containing the bidentate consensus motif pYXXZ-X_n-pYXXZ, derived from a potential Syk tyrosine kinase interaction site within the CD3ε-chain, binds in a canonical fashion aligned roughly perpendicular to the central β-sheet within each SH2 domain in a head-to-tail manner — the N-terminal pYXXZ motif (N-pYXXZ) interacts with the C-SH2 and the C-terminal pYXXZ motif (C-pYXXZ) with the N-SH2 (Figure 2d). The inter-motif region comprised of a variable sequence of seven residues takes a nearly complete helical turn at halfway close to the inteface of the tandem domains, giving the peptide an overall L-shaped conformation. Within each SH2 domain of Syk, lie a basic and an hydrophobic pocket for binding pY and pY(+3) within each of the two motifs, respectively. The pY within each motif is coordinated by 2-3 arginines, while the pY(+3) residue becomes deeply buried in the hydrophobic pockets created by well-conserved leucine, isoleucine and tyrosine residues. The ZAP70 tyrosine kinase also interacts with bidentate tyrosine phosphorylated peptides in a similar manner but with one important difference [7]. While the binding pockets for pY in both N-SH2 and C-SH2 domains of Syk act independently of each other, the pY binding pocket for the C-pYXXZ motif within the N-SH2 of ZAP70 requires residues from both domains — residues from one domain are needed to create the pY binding pocket in the other. In the case of tandem SH2 domains of SHP2 tyrosine phosphatase [11], the tyrosine phosphorylated peptides bind to distinct pockets within each SH2 domain although the orientation of peptide binding is different to what is observed in Syk [8] and ZAP70 [7] tyrosine kinases.

In short, the tandem SH2 domains underlie the hallmark of tandem domain function. The ability of each domain to restrict the freedom of the other not only optimizes the spatial organization of both domains but also makes them more attractive towards their bidentate ligands. The increase in affinity of several orders of magnitude toward their bidentate ligands clearly results from favorable enthalpic and entropic factors acquired by such organization of tandem SH2 domains.

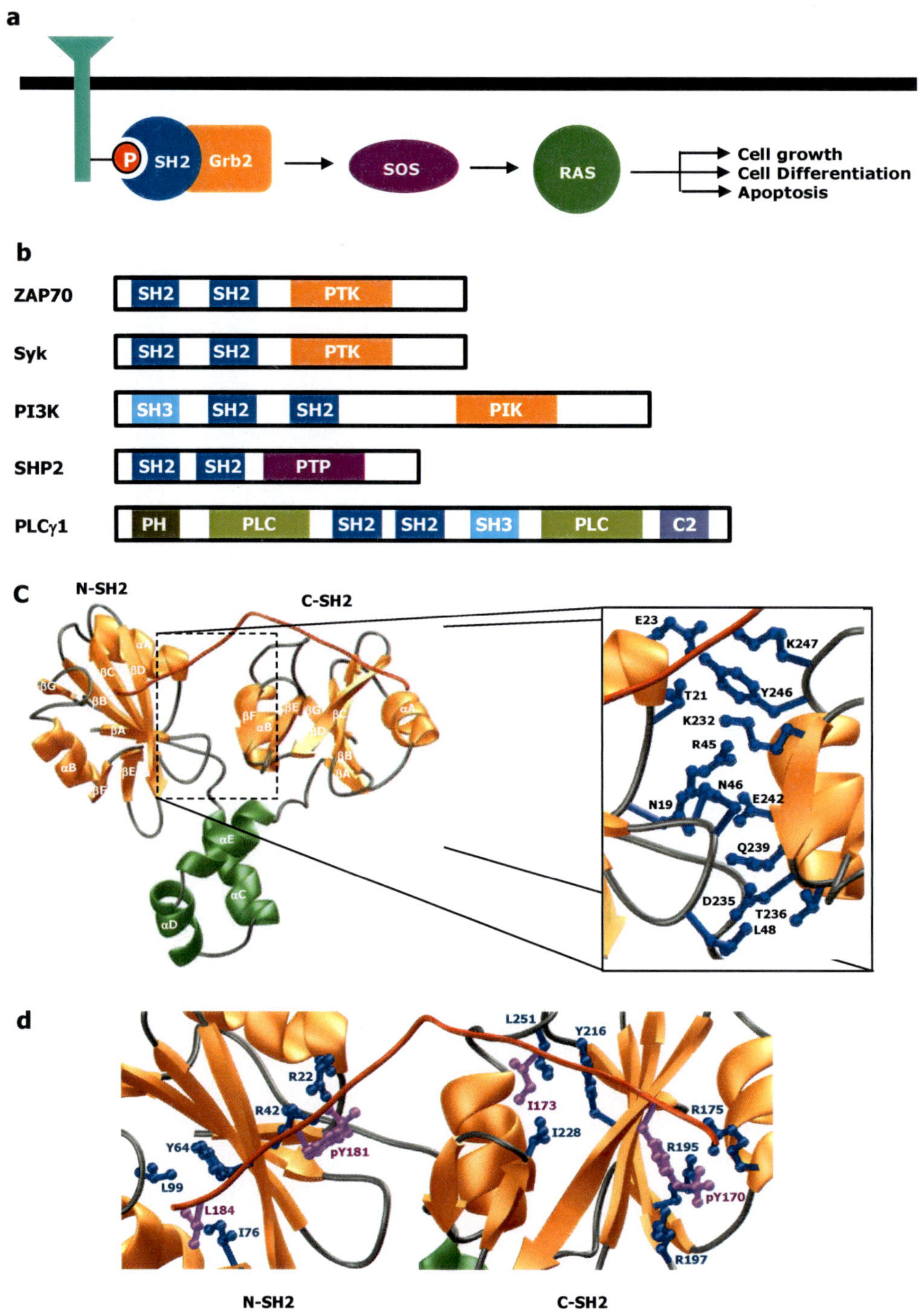

Figure 2. Tandem SH2 domains. (a) A schematic showing the involvement of the SH2 domain in signal transduction. The SH2 domain specifically recognizes pY and residues C-terminal to pY in cytoplasmic proteins including the cytoplasmic tails of cell surface receptors. Upon binding to pY, the SH2-containing proteins recruit SOS and RAS, which in turn can initiate signaling pathways leading to cellular processes such as cell growth, cell differentiation and apoptosis. (b) Examples of some proteins containing tandem repeats of SH2 domains. (c) Crystal structure of the tandem SH2 domains of the

protein tyrosine kinase Syk. The N-terminal and the C-terminal domains are respectively labeled N-SH2 and C-SH2. The secondary structural elements of the two domains are shown in yellow, the helices αC, αD and αE that constitute the linker stem region are colored green and the bidentate peptide ligand is shown in red. The inset shows the interdomain interactions between the sidechains of key residues from each domain. (d) Top view of the tandem SH2 domains shown in (c) providing a closer look at the peptide binding sites within each domain

Tandem WW Domains

The WW domain was discovered in the mid 1990s and named after the presence of two highly conserved tryptophan residues spaced approximately 20 residues apart [40, 41]. The WW domain is present in a wide variety of eukaryotic signaling and structural proteins [41, 42]. The WW domain allows proteins to selectively bind their biological partners through the recognition of proline-rich sequences in a manner similar to the SH3 domain. However, some WW domains may employ a novel mechanism in which the recognition of proline-rich sequences is tolerated. Thus, for example, the WW domains of the suppressor of deltex Su(dx) recognize Notch, which does not contain a typical proline-rich motif recognized by other WW domains (Figure 3a). The WW and the SH3 domains are believed to be evolutionarily related but have diverged in their roles to meet the ever increasing demands of cellular signaling and specificity in higher organisms [43].

The WW domain is one of the best-characterized protein modules both structurally and functionally. Such analysis has led to the sub-calssification of WW domains into four distinct categories on the basis of their distinct ligand binding requirements. A number of excellent reviews have given the WW domain a comprehensive coverage [43, 44]. Our focus here is to shed light on the newly discovered feature of WW domains: their presence as tandem repeats in a wide variety of cellular proteins (Figure 3b). In an attempt to understand the molecular mechanism of how such tandem repeats function, structures of tandem WW domains of the yeast splicing factor Prp40 [45] and the suppressor of deltex Su(dx) [46] have recently been determined to high resolution. These tandem WW domains share a number of common features but there are also differences that indicate that WW domains may have evolved unique mechanisms to operate in tandem. Thus, while the triple-stranded β-sheets of tandem domains of both the Prp40 and Su(dx) are held together by a linker, the flexibility of this region seems to hold clues to understanding how they may act in a co-operative manner.

In Prp40, the linker region is comprised of a well-ordered helix that appears to impart strict rigidity and a fixed orientation upon its tandem domains WW1 and WW2 (Figure 3c). The two domains essentially act as a single rigid body and contain putative hydrophobic pockets on opposite faces with characteristic features for binding to proline-rich sequences. In WW1, the putative binding pocket is comprised of hydrophobic residues such as Y15, Y17 and W26 clustered on one face of the β-stranded structure. In WW2, the equivalent residues are Y56, Y58 and W67. Interestingly, the sides of the two β-stranded structures facing each other also contain a highly conserved patch of hydrophobic residues such as W4, I14 and Y16 in WW1 and W45, V55 and Y57 in WW2. These residues have the potential to interact in an inter-molecular fashion and thus bring the two WW domains closer together in space.

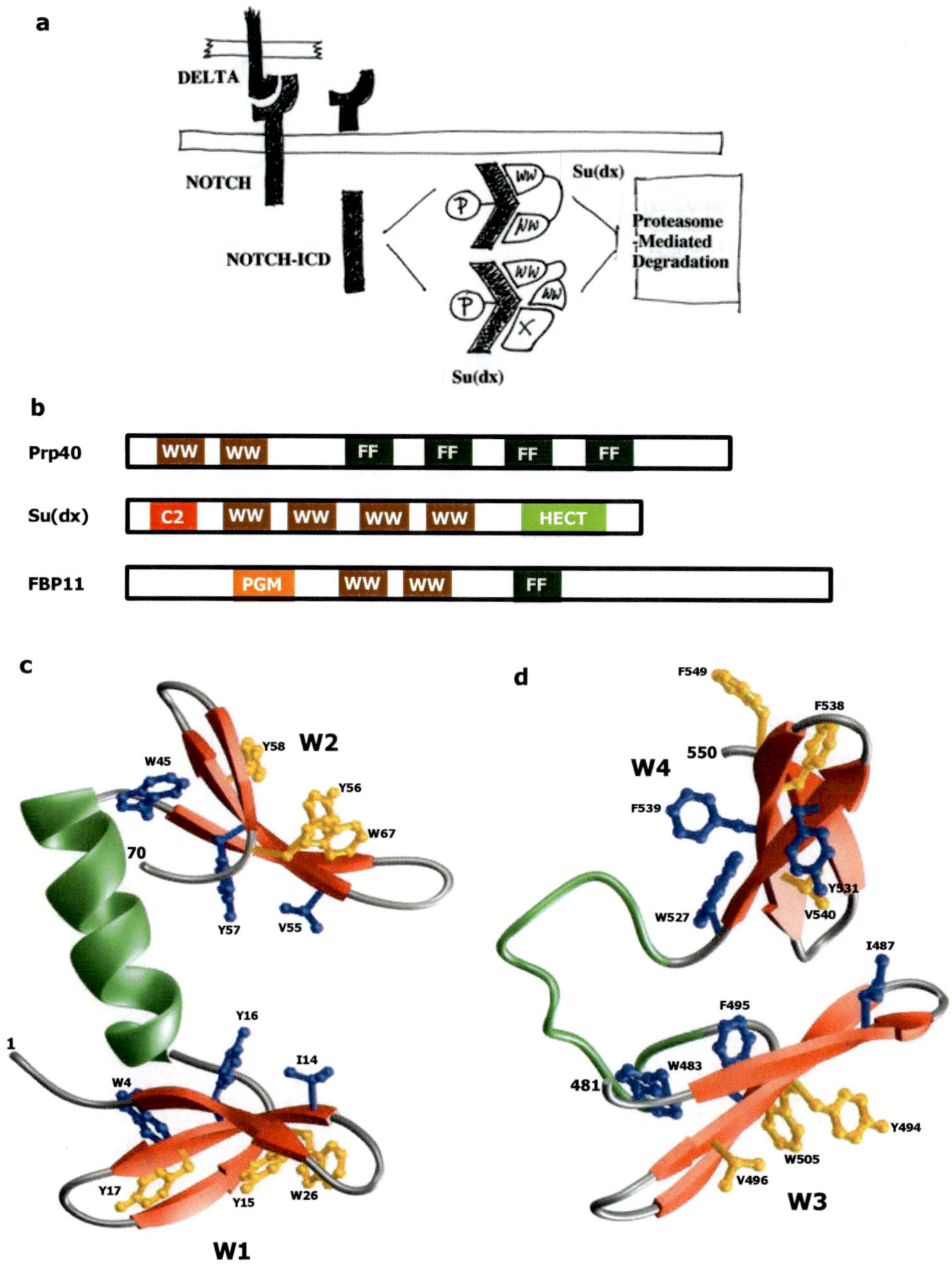

Figure 3. WW domains. (a) A schematic showing the involvement of the WW domains of the suppressor of deltex Su(dx) in the proteasome-mediated degradation of Notch. Activation of Notch upon Delta ligand binding is accompanied by proteolytic processing that releases an intracellular domain of Notch (Notch-ICD) from the membrane to cytoplasm and ultimately to the nucleus, where Notch regulates transcription. The metabolism of Notch-ICD is controlled by phosphorylation that initiates the ubiquitination of Notch by the E3 ubiquitin ligase suppressor of deltex Su(dx). The interaction between Notch-ICD and Su(dx) could be direct. Although Notch protein does not have a typical –PPXY- motif needed for binding to Su(dx) WW domains, the tandem W3 and W4 of Su(dx) could recognize non-linear peptides or mimotopes on Notch. X depicts a putative protein that is part of the tripartite complex targeting Notch for proteasome-mediated degradation. X could contain a perfect –

PpXY- motif. (b) Examples of some proteins containing tandem arrays of WW domains. (c) Solution structure of the tandem domains WW1-WW2 of the yeast splicing factor Prp40. (d) Solution structure of the tandem domains WW3-WW4 of suppressor of deltex Su(dx). In both (c) and (d), the tandem WW domains are shown in red, the interdomain linker is in green, the sidechains of key residues constituting the putative binding pocket for proline-rich sequences are colored yellow, and sidechains of key residues constituting the hydrophobic surface on the opposite sides to the ligand binding pocket are depicted in blue.

However, such interactions are not observed and the two domains are merely held together and maintained in a fixed orientation by the interdomain helix. Such organization of WW domains not only renders them capable of binding to distinct sites in target proteins but also fits well with the role of Prp40 in its ability to interact simultaneously and bridge precisely between target sites within the splicing machinery. In contrast, the linker region between the WW3 and WW4 tandem domains of Su(dx) does not adopt helical conformation and is largely comprised of a rather flexible loop of approximately 20 residues in length (Figure 3d). The flexibility of this interdomain loop is believed to be critical to the function of the two WW domains in Su(dx). However, despite such flexibility of the linker loop, the two domains behave as one rigid body and their orientation, although not as rigid as observed with the WW domains of Prp40, relative to each other is also somewhat restricted. Like the tandem WW domains of Prp40, the tandem WW domains of Su(dx) are positioned in such a fashion so as to allow them to interact with their target sites independent of each other. In WW3, the putative binding pocket for proline-rich sequences is comprised of hydrophobic residues such as Y494, V496 and W505 clusterd on one face of the triple-stranded β-sheet structure. The equivalent residues in WW4 are F538, V540 and F549, with the latter residue mimicking the role of the second conserved tryptophan found in WW3. The substitution of phenylalanine for the second conserved tryptophan does not necessarily render WW4 domain functionally distinct from classical WW domains in which both tryptophans are conserved. In support of this view is the observation that although the second tryptophan in the second WW domain of the tandem WW domains of the tight junction protein MAGI1 is replaced by a tyrosine, both WW domains interact with the proline-rich motif –PPXY- found in synaptotagmins [47]. The sides of WW3 and WW4 facing each other also contain highly conserved patches of hydrophobic residues such as W483, I487 and F495 in WW3, and W527, Y531 and F539 in WW4. The orientation of the interdomain linker however prevents these residues to come in close proximity of each other and engage in intermolecular interactions.

Thus, like the tandem WW domains of Prp40, the tandem WW domains of Su(dx) are unable to interact with each other even though they are tethered together in a more or less fixed orientation required for their biological function. Unlike the tandem WW domains of Prp40, which seem to undergo little or negligible structural change upon ligand binding, the tandem WW domains of Su(dx) are partially unstructured in unliganded state and attain full native globular-like structure only upon binding to their cognate ligands. It is believed that the binding of ligand to WW3 not only triggers a conformational change within this domain but that it is also transmitted to WW4 through the flexible interdomain loop such that the latter in turn adopts a conformation optimal for ligand binding and may return the compliments by further stabilizing WW3. In this manner, ligand binding to one domain synergistically enhances the ligand binding potential of the other. Worthy of note is also the fact that in Prp40, both WW domains are very similar to each other including the core of central aromatics. In Su(dx), however, the WW3 and WW4 tandem domains are significantly different in their binding pockets suggesting a possibility that the same set of two domains may act co-operatively by binding diverse sequences on a single substrate or that they interact with a variety of different substrates to be targeted for degradation.

In summary, the arrangement of WW domains of Prp40 and Su(dx) in a tandem fashion appears to present an important facet of their biology. In the case of Prp40, the tandem WW domains are held together in a fixed orientation such that their ligand binding pockets face to the outside and away from each other. Such a modular organization speaks volumes about their ability to participate and simultaneously act as a correctly set "molecular vice" that positions splicing factors to engage in RNA processing between splicing factors that constitute the splicing machinery. The need for the WW domains of Su(dx) to come together as a single functioning unit seems to be of a very different origin. These domains appear to take up their native structure only upon interaction with their cognate ligands and, as a result, they have adopted a synergistic mechanism in which ligand binding to one domain enhances the ability of the other to interact with its target sites and vice versa.

Tandem PDZ Domains

PDZ domain, a widely distributed protein interaction module among eukaryotic proteins, is involved in important cellular processes such as organization of signal transduction complexes, clustering membrane receptors and maintaining cell polarities [48-50]. PDZ domains usually bind C-terminal sequences, with no well-defined consensus sequence, of potential targets but they may also heterodimerize to allow PDZ domains from two different proteins to act as an adaptor to couple signal between different components [51-53]. PDZ domains are often arranged in duplets, triplets or multiplets in a wide range of proteins including PSD95 [13], syntenin [14], X11 [16] and GRIP [18] (Figure 4a).

As expected from such a repetitive modular arrangement, the PDZ domains interact with their ligands in a cooperative manner and, in many cases, the individual domains are either unfolded or do not at all bind the ligand. The tandem repeat, comprised of the fourth and fifth PDZ domains (PDZ4 and PDZ5), in GRIP go a step further. Here, the two PDZ domains act as a single supramodule unit, in which only one of the two PDZ domains (PDZ5) actually interacts with the C-terminal tail of the glutamate receptor (GluR) [54], while the other is merely required to stabilize it (Figure 4b). In an attempt to understand the molecular basis of this interaction, Zhang and co-workers solved the solution structure of the tandem PDZ4 and PDZ5 domains of GRIP critical for glutamate receptor binding [17]. In the tandem repeat, each PDZ domain is constructed on a compact globular fold composed of a six-stranded (βA-βF) anti-parallel β-barrel capped at each end with an α-helix (αA and αB) (Figure 4c) — a topology characteristic of a canonical PDZ domain [55]. The two domains are tightly held together in a fixed orientation by a five-residue sequence at the N-terminus of PDZ4 and a ten-residue interdomain linker region between PDZ4 and PDZ5. While the N-terminal half of the linker sequence interacts with the N-terminal extension of PDZ4, the C-terminal half of the linker sequence forms an anti-parallel β-sheet structure with the first β-strand (βA) of the β-barrel in PDZ5. Such a tight network of interactions generated by the N-terminal extension and the linker region, as well as due to numerous additional interactions between the two domains results in close packing of PDZ4 and PDZ5 with a restrained orientation (Figure 4c, inset). Interestingly, while PDZ5 completely unfolds in isolation, the individual PDZ4, although less stable in isolation than within the PDZ4-PDZ5 tandem repeat, retains essentially the same fold as that observed in the structure of the tandem repeat PDZ4-PDZ5 — except for the N-terminal extension and the linker region which are unstructured and exhibit high flexibility relative to their conformations in the tandem repeat PDZ4-PDZ5 [17]. The highly conserved nature of residues within the N-terminal extension and the linker sequence in all members of the GRIP family together with the structural data argues strongly in favor of the critical role of these regions in the stabilization of PDZ5 by PDZ4.

To further delineate the role of PDZ4 in the interaction of GRIP through its PDZ5 domain with the glutamate receptor, Zhang and co-workers performed NMR titration of the tandem repeat PDZ4-PDZ5 with a peptide derived from the tail of the glutamate receptor [17]. The salient observation that virtually no residues in PDZ4, the N-terminal extension and the linker region underwent perturbation upon the addition of the peptide further confirms the view that only PDZ5 directly binds the glutamate receptor and that the role of PDZ4 and the terminal regions is merely to chaperone PDZ5. A vast majority of perturbations, however, correspond to residues within the PDZ5 domain, particularly in the αB-βB groove that represents the potential peptide binding site. To add insult to injury, the amino acid sequence of PDZ4 reveals that it is a canonical PDZ domain and contains the αB-βB groove for interaction with a peptide ligand — which inserts into this groove as an additional β-strand. The structure of PDZ4, both in isolation and within the tandem repeat PDZ4-PDZ5, however shows that the αB helix is tilted ~52° outwards with respect to the αB helix in PDZ5. Such a tilt of αB helix in PDZ4 results in the narrowing of the αB-βB groove in PDZ4, rendering it unable to accommodate the incoming peptide (Figure 4d). In contrast, the αB-βB groove in a canonical PDZ domain such as PDZ5 assumes a rather open conformation ideally suited for peptide binding.

In conclusion, the PDZ story tells us that tandem domains may not always act in a mutually beneficial way but that one domain may sacrifice its function for the survival of the other. This co-operative action of the PDZ4 and PDZ5 tandem domains appears to resemble the role of worker slave toward its queen master in a beehive and demonstrates the very subtle behavior of protein modules.

Thermodynamic Considerations

In thermodynamic terms, the binding of ligand to a macromolecule is best rationalized by consideration of the contribution of the overall change in enthalpy (ΔH) and entropy (ΔS) of the system to the free energy of binding (ΔG). The relationship between these classical thermodynamics parameters is given by the equation:

$$\Delta G = \Delta H - \Delta S$$

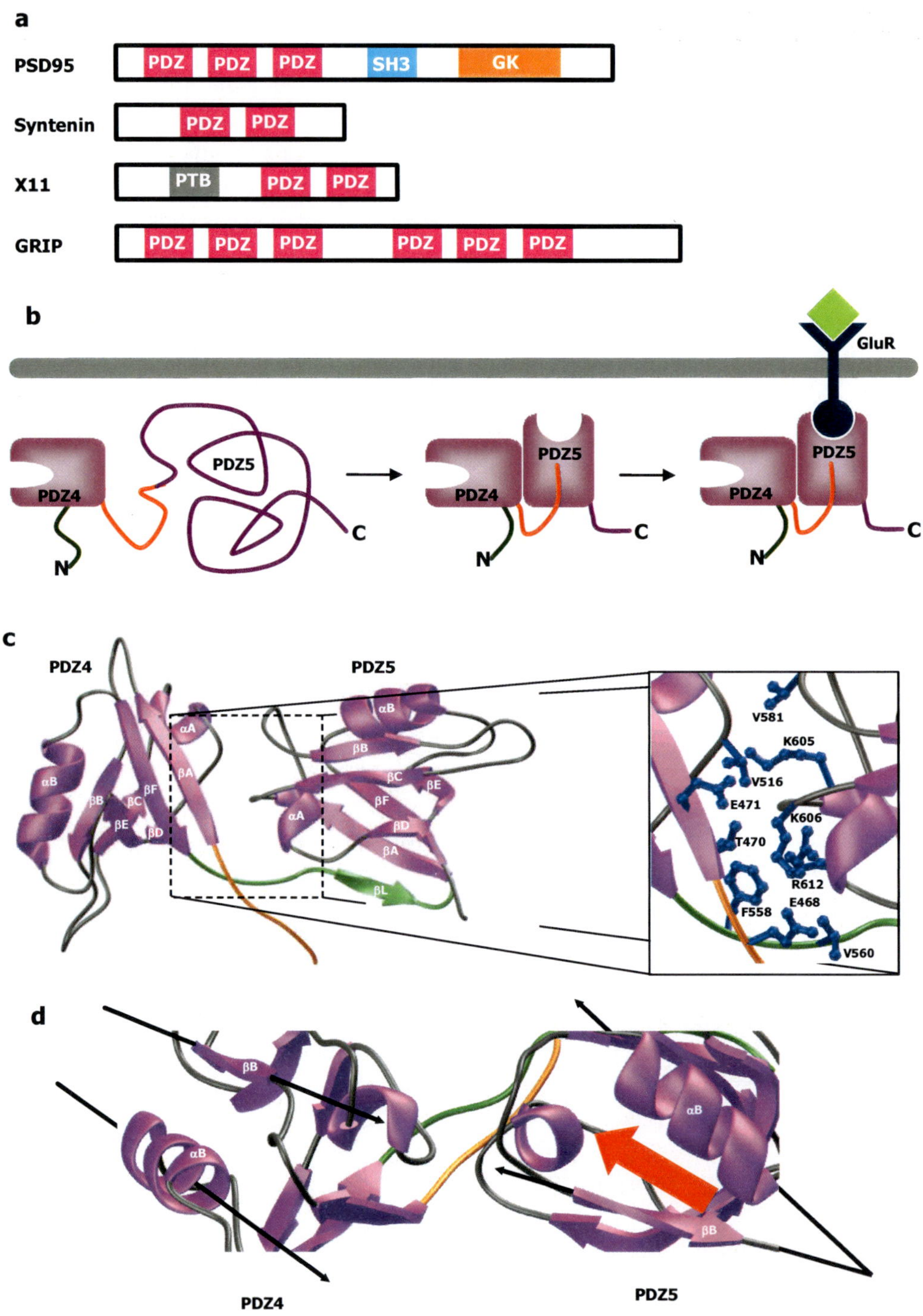

Figure 4. Tandem PDZ domains. (a) Examples of some proteins containing tandem repeats of PDZ domains. (b) A simplified model for the interaction of PDZ4-PD5 tandem repeat with the glutamate receptor (GluR) tail. The stabilization of the tertiary structure of PDZ5 by PDZ4, and in particular by the N-terminal extension (green) and the interdomain linker sequence (red), results in the formation of a supramodule PDZ4-PDZ5. The supramodule PDZ4-PDZ5 acts as a single functional unit in which

PDZ5 directly binds the GluR tail, while PDZ4 merely provides the supporting role and is unable to accommodate the GluR tail in its rather narrow and deformed binding pocket. (c) Solution structure of the tandem PDZ4-PDZ5 domains of GRIP. The secondary structural elements of the two domains are shown in purple, the N-terminal region of PDZ4 in yellow and the interdomain linker region containing the strand βL in green. The inset shows the interdomain interactions between the sidechains of key residues from each domain. (d) Top view of the tandem PDZ4-PDZ5 domains shown in (c) providing a closer look at the peptide binding sites within each domain. Thick black arrow-headed lines are drawn through the axis of αB and βB, that constitute the binding groove, to indicate the orientation of helix αB relative to the strand βB. The position of peptide ligand in PDZ5 is indicated by a red arrow.

The ΔG of the interaction between two partners not only drives the binding process to completion but is also directly related to the binding or dissociation constant (K_d) by the equation:

$$\Delta G = RT\ln K_d$$

Where R is the universal molar gas constant and T is the absolute temperature. The K_d describes the binding affinity of the ligand to a macromolecule. The relationship between K_d and binding affinity is a reciprocal one — the lower the value of the K_d in the units of concentration, the higher the binding affinity. From the equations above, we can see how contributions from ΔH and ΔS, two parameters that can be easily rationalized in terms of structural data, would directly affect the binding affinity:

$$K_d = \exp[(\Delta H - \Delta S)/RT]$$

The foregoing argument implies that the higher binding affinity (lower K_d) of tandem domains to their cognate ligands must be the result of either an increase in ΔH (more negative) or ΔS (more positive) or both. From the structures of tandem domains discussed above (Figures 1-3), it does not appear how ligand binding to tandem domains could result in higher ΔH relative to individual domains. Although both domains within the tandem repeat act in a synergistic manner toward ligand binding, the binding pocket within each domain is distinct and acts more or less independently – the residues lining the binding pockets are essentially the same in tandem domains as in individual domains. Thus ΔH resulting largely from the electrostatic and hydrophobic interactions formed by these residues must be more or less the same for ligand binding to individual domain versus tandem domains. The above argument leads us to speculate that ΔS must be the driving force for an enhancement of ligand binding in tandem domains relative to individual domains. How can we rationalize this? The tandem domains are locked into a fixed orientation via tight interfacial interactions as discussed above (Figures 1-4). As a consequence, the freedom of each domain relative to the other is restricted. In contrast, the individual domains experience no such restriction of freedom. Even though ligand binding would be expected to result in an overall decrease in ΔS in each case, the value of ΔS is likely to be much less negative in the case of ligand binding to tandem domains versus individual domains for the latter would experience an overall greater loss of freedom compared to the former which had limited freedom to begin with. We make use of binding funnels and free energy profiles to further emphasize the entropic advantages involved in ligand binding to tandem domains compared to individual domains (Figure 5). The binding funnels show how ligand binding to tandem domains results in smaller ΔS, while the free energy profiles show that such a decrease in ΔS results in an overall higher ΔG for ligand binding to tandem domains relative to individual domains (Figures 5a and 5b). This increase in ΔG could account for the enhanced ligand binding to tandem domains versus the individual domains.

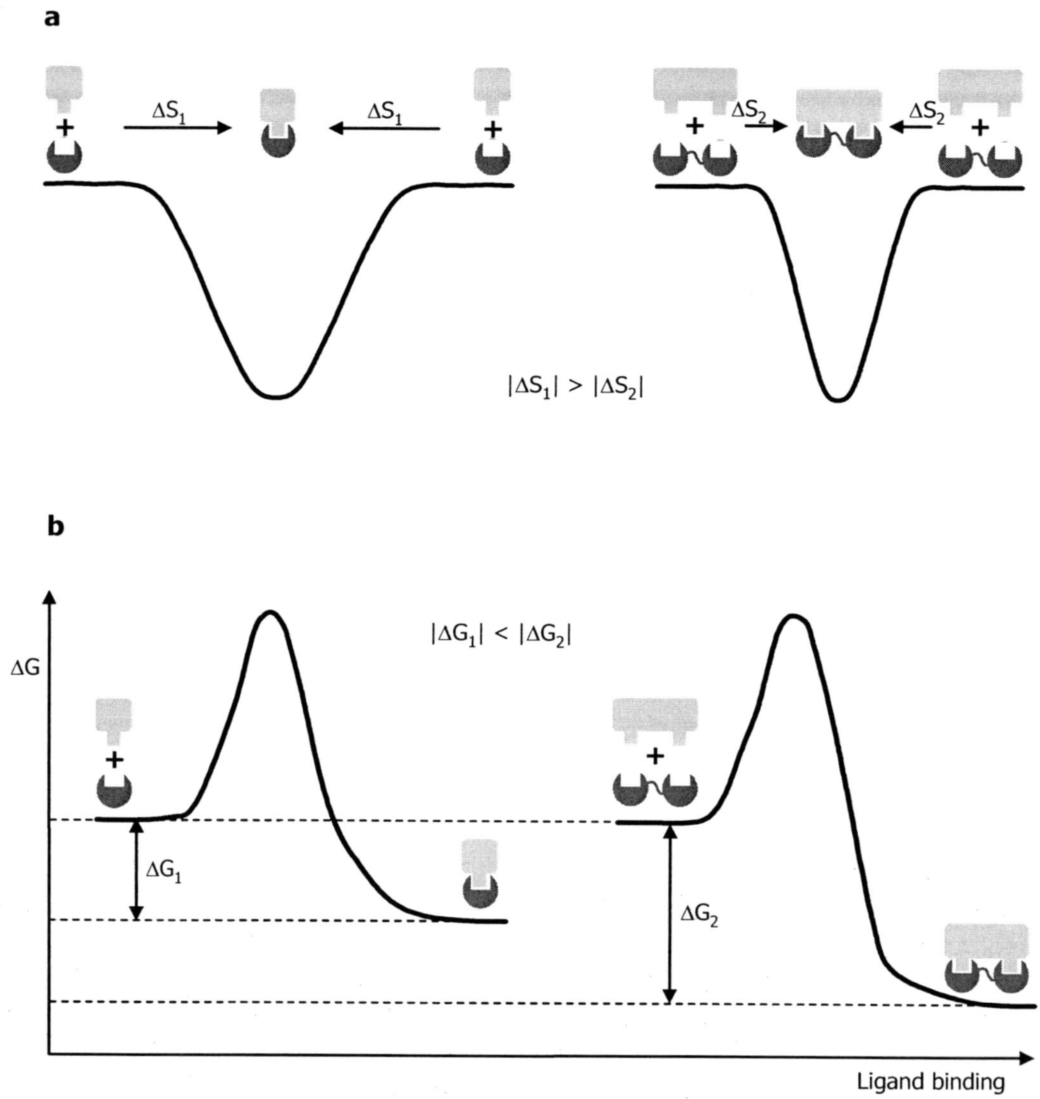

Figure 5. Thermodynamics of ligand binding to tandem domains. (a) Binding funnels for the binding of ligand to an individual domain (left) versus the binding of a bidentate ligand to tandem domains (right). The vertical axis of the funnels represents the binding enthalpy (ΔH) while the many lateral axis denote the conformational entropy (ΔS). (b) Free energy (ΔG) profiles for the binding of a ligand to an individual domain (left) versus the binding of a bidentate ligand to tandem domains (right)

Future Perspectives

The occurrence of tandem repeats across a wide range of signaling proteins seems to be a mechanism nature has adopted to improve the efficiency and fidelity of cellular signal transduction. Such a mechanism is elegantly demonstrated by the tandem bromodomains [6], the tandem SH2 domains [7, 8], the tandem WW domains [45, 46] and the tandem PDZ domains [13, 15, 17]. Although the scope of our current review has limited us to focus only on these four modules, that are widely encountered across a diverse spectrum of cellular proteins, the use of tandem repeats to carry out biological functions extends to many other protein modules and new examples of tandem repeats are being discovered as we write. Importantly, the regulation of the activity of receptor tyrosine phosphatases (rPTPs) through the involvement of tandem repeats has been known for over a decade [56]. The breast cancer associated protein BCRA1 has been shown to recognize its phosphorylated targets through its C-terminal BRCT tandem domains [57]. Other well-documented examples of protein modules acting in tandem repeats to attain co-operativity toward their targets include the tandem SH3 domains of NADPH oxidase [58], the tandem GAF domains of bacterial adenylyl cyclase [59], the tandem C2 domains of synaptotagmin involved in the engagement of SNAREs [60] and the tandem FF domains found in proteins associated with RNA splicing [61]. Extensive structural analysis of tandem domains over the last decade or so has begun to shed light on the molecular mechanism of their co-operative action. Yet, we have a long way to go before we can fully understand the details of such co-operativity. First, the structures of tandem domains in complex with their bidentate ligands are clearly warranted. Understanding the molecular basis of co-operativity observed in tandem domains in thermodynamic and kinetic terms is also an important facet of the molecular picture. Direct measurements on ligand binding to tandem versus individual domains are thus needed. The design of proteomic arrays of modular domains (SH2, PDZ, SH3 and WW domain arrays are already available from Panomics) should contain tandems, and double tandems (especially for WW, FF and PDZ) and also engineered permutations of tandem domains. Such proteomic arrays would significantly help in the understanding of functional co-evolution of closely located domains and could provide new molecular tools for re-directing signaling events for therapeutic purposes. A combined venture between academia and industry will undoubtedly provide further insights into how tandem domains work at molecular level and may pave the way for the development of a new generation of drugs aimed at exploiting the unique features employed by tandem domains. The ever increasing myriad of tandem repeats clearly poses a new challenge in understanding cellular signal transduction at a new level and there is indeed a lot at stake.

References

[1] Sadowski, I., Stone, J.C., and Pawson, T. (1986). A noncatalytic domain conserved among cytoplasmic protein-tyrosine kinases modifies the kinase function and transforming activity of Fujinami sarcoma virus P130gag-fps. *Mol. Cell. Biol. 6*, 4396-4408.

[2] Pawson, T. (1995). Protein modules and signalling networks. *Nature 373*, 573-580.

[3] Pawson, T., and Saxton, T.M. (1999). Signaling Networks—Do All Roads Lead to the Same Genes? *Cell 97*, 675-678.

[4] Pawson, T., and Scott, J.D. (1997). Signaling through scaffold, anchoring, and adaptor proteins. *Nature 278*, 2075-2080.

[5] Pawson, T., and Nash, P. (2000). Protein-protein interactions define specificity in signal transduction. *Genes Dev. 14*, 1027-1047.

[6] Jacobson, R.H., Ladurner, A.G., King, D.S., and Tjian, R. (2000). Structure and Function of a Human TAFII250 Double Bromodomain Module. *Science 288*, 1422-1425.

[7] Hatada, M.H., and al, e. (1995). Molecular Basis for Interaction of the Protein Tyrosine Kinase ZAP70 With the T-Cell Receptor. *Nature 377*, 32-38.

[8] Fütterer, K., Wong, J., Grucza, R.A., Chan, A.C., and Waksman, G. (1998). Structural basis for syk tyrosine kinase ubiquity in signal transduction pathways revealed by the crystal structure of its regulatory SH2 domains bound to a dually phosphorylated ITAM peptide. *J. Mol. Biol. 281*, 523-537.

[9] Carpenter, C.L., Auger, K.R., Chanudhuri, M., Yoakim, M., Schaffhausen, B., Shoelson, S., and Cantley, L.C. (1993). Phosphoinositide 3-kinase is activated by phosphopeptides that bind to the SH2 domains of the 85-kDa subun. *J. Biol. Chem. 268*, 9478-9483.

[10] Hof, P., Pluskey, S., Dhe-Paganon, S., Eck, M.J., and Shoelson, S.E. (1998). Crystal structure of the tyrosine phosphatase SHP-2. *Cell 92*, 441-450.

[11] Eck, M.J., Pluskey, S., Trub, T., Harrison, S.C., and Shoelson, S.E. (1996). Spatial Constraints on the Recognition of Phosphoproteins by the tandem SH2 Domains of the Phosphatase SH-PTP2. *Nature 379*, 277-280.

[12] Rhee, S.G., and Bae, Y.S. (1997). Regulation of Phosphoinositide-specific Phospholipase C Isozymes. *J. Biol. Chem. 272*, 15045-15048.

[13] Long, J.-F., Tochio, H., Wang, P., Fan, J.-S., Sala, C., Niethammer, M., Sheng, M., and Zhang, M. (2003). Supramodular structure and synergistic target binding of the N-terminal tandem PDZ domains of PSD-95. *J. Mol. Biol. 327*, 203-214.

[14] Zimmermann, P., Tomatis, D., Rosas, M., Grootjans, J., Leenaerts, I., Degeest, G., Reekmans, G., Coomans, C., and David, G. (2001). Characterization of Syntenin, a Syndecan-binding PDZ Protein, as a Component of Cell Adhesion Sites and Microfilaments. *Mol. Biol. Cell 12*, 339-350.

[15] Kang, B.S., Cooper, D.R., Jelen, F., Devedjiev, Y., Derewenda, U., Dauter, Z., Otlewski, J., and Derewenda, Z.S. (2003). PDZ Tandem of Human Syntenin: Crystal Structure and Functional Properties. *Structure 11*, 459-468.

[16] Lau, K.-F., McLoughlin, D.M., Standen, C., and Miller, C.C.J. (2000). X11 and X11 Interact with Presenilin-1 via Their PDZ Domains. *Mol. Cell. Neurosci. 16*, 557-565.

[17] Feng, W., Shi, Y., Li, M., and Zhang, M. (2003). Tandem PDZ repeats in glutamate receptor–interacting proteins have a novel mode of PDZ domain–mediated target binding. *Nat. Struct. Biol. 10*, 972-978.

[18] Zhang, Q., Fan, J.-S., and Zhang, M. (2001). Interdomain Chaperoning between PSD-95, Dlg, and Zo-1 (PDZ) Domains of Glutamate Receptor-interacting Proteins. *J. Biol. Chem. 276*, 43216-43220.

[19] Harris, T.K. (2003). PDK1 and PKB/Akt: ideal targets for development of new strategies to structure-based drug design. *IUBMB Life 55*, 117-126.

[20] Strahl, B.D., and Allis, C.D. (2000). The language of covalent histone modifications. *Nature 403*, 41-45.

[21] Fischle, W., Wang, Y., and Allis, C.D. (2003). Binary switches and modification cassettes in histone biology and beyond. *Nature 425*, 475-479.

[22] Jenuwein, T., and Allis, C.D. (2001). Translating the histone code. *Science 293*, 1074-1080.

[23] Turner, B.M. (2002). Cellular Memory and the Histone Code. *Cell 111*, 285-291.

[24] Zeng, L., and Zhou, M.-M. (2001). Bromodomain: An Acetyl-Lysine Binding Domain. *FEBS Lett 513*, 124-128.

[25] Dhalluin, C., Carlson, J.E., Zeng, L., He, C., Aggarwal, A.K., and Zhou, M.-M. (1999). Structure and ligand of a histone acetyltransferase bromodomain. *Nature 399*, 491-496.

[26] Hudson, B.P., Martinez-Yamout, M.A., Dyson, H.J., and Wright, P.E. (2000). Solution structure and acetyl-lysine binding activity of the GCN5 bromodomain. *J. Mol. Biol. 304*, 355-370.

[27] Owen, D.J., Ornaghi, P., Yang, J.C., Lowe, N., Evans, P.R., Ballario, P., Neuhaus, D., Eiletici, P., and Travers, A.A. (2000). The structural basis for the recognition of acetylated histone H4 by the bromodomain of histone acetyltransferase gcn5p. *EMBO J. 19*, 6141-6149.

[28] Rhee, K., Brunori, M., Besset, V., Trousdale, R., and Wolgemuth, D.J. (1998). Expression and potential role of Fsrg1, a murine bromodomain-containing homologue of the Drosophila gene female sterile homeotic. *J. Cell Sci. 111*, 3541-3550.

[29] Maruyama, T., Farina, A., Dey, A., Cheong, J., Bermudez, V.P., Tamura, T., Sciortino, S., Shuman, J., Hurwitz, J., and Ozato, K. (2002). A Mammalian Bromodomain Protein, Brd4, Interacts with Replication Factor C and Inhibits Progression to S Phase. *Mol. Cell. Biol. 22*, 6509-6520.

[30] Haynes, S.R., Mozer, B.A., Bhatia-Dey, N., and Dawid, I.B. (1989). The Drosophila fsh locus, a maternal effect homeotic gene, encodes apparent membrane proteins. *Dev. Biol. 134*, 246-257.

[31] Ladurner, A.G., Inouye, C., Jain, R., and Tjian, R. (2003). Bromodomains mediate an acetyl-histone encoded antisilencing function at heterochromatin boundaries. *Mol. Cell 11*, 365-376.

[32] Matangkasombut, O., and Buratowski, S. (2003). Different sensitivities of bromodomain factors 1 and 2 to histone H4 acetylation. *Mol. Cell 11*, 353-363.

[33] Pawson, T., and Nash, P. (2003). Assembly of Cell Regulatory Systems Through Protein Interaction Domains. *Science 300*, 445-452.

[34] Pelicci, G., Lanfrancone, L., Grignani, F., McGlade, J., Cavallo, F., Forni, G., Nicoletti, I., Grignani, F., and Pawson, T. (1992). A novel transforming protein (Shc) with an SH2 domain is implicated in mitogenic signal transduction. *Cell 70*, 93-104.

[35] Salcini, A., McGlade, J., Pelicci, G., Nicoletti, I., Pawson, T., and Pelicci, P. (1994). Formation of Shc-Grb2 complexes is necessary to induce neoplastic transformation by overexpression of Shc proteins. *Oncogene 9*, 2827-2836.

[36] Rozakis-Adcock, M., McGlade, J., Mbamalu, G., Pelicci, G., Daly, R., Li, W., Batzer, A., Thoma, S., Brugge, J., Pelicci, P.G., Schlessinger, J., and Pawson, T. (1992). Association of the Shc and Grb2/Sem5 SH2-containing proteins is implicated in activation of the ras pathway by tyrosine kinases. *Nature 360*, 689-692.

[37] Reth, M. (1989). Antigen Receptor Tail Clue. *Nature 338*, 383-385.

[38] Kurosaki, T. (1997). Molecular Mechanisms in B Cell Antigen Receptor Signaling. *Curr. Opin. Immunol. 9*, 309-318.

[39] Waksman, G., Kominos, D., Robertson, S.C., Pant, N., Baltimore, D., Birge, R.B., Cowburn, D., Hanafusa, H., Mayer, B.J., Overduin, M., Resh, M.D., Rois, C.B., Silverman, L., and Kuriyan, J. (1992). Crystal structure of the phosphotyrosine recognition domain SH2 of v-src complexed with tyrosine phosphorylated peptides. *Nature 358*, 646-653.

[40] Bork, P., and Sudol, M. (1994). The WW domain: a signalling site in dystrophin? *Trends Biochem Sci 19*, 531-533.

[41] Sudol, M., Chen, H., Bougeret, C., Einbond, A., and Bork, P. (1995). Characterization of a novel protein-binding module -- the WW domain. *FEBS Lett 369*, 67-71.

[42] Hu, H., Columbus, J., Zhang, Y., Wu, D., Lian, L., Yang, S., Goodwin, J., Luczak, C., Carter, M., Chen, L., James, M., Davis, R., Sudol, M., Rodwell, J., and Herrero, J.J. (2004). A map of WW domain family interactions. *Proteomics 4*, 643-655.

[43] Macias, M.J., Wiesner, S., and Sudol, M. (2002). WW and SH3 domains, two different scaffolds to recognize proline-rich ligands. *FEBS Lett 513*, 30-37.

[44] Kato, Y., Nagata, K., Takahashi, M., Lian, L., Herrero, J.J., Sudol, M., and Tanokura, M. (2004). Common mechanism of ligand recognition by group II/III WW domains: redefining their functional classification. *J. Biol. Chem. 279*, 31833-31841.

[45] Wiesner, S., Stier, G., Sattler, M., and Macias, M.J. (2002). Solution Structure and Ligad Recognition of the WW Domain of the Yeast Splicing Factor Prp40. *J . Mol. Biol.*, 807-822.

[46] Fedoroff, O.Y., Townson, S.A., Golovanov, A.P., Baron, M., and Avis, J.M. (2004). The Structure and Dynamics of Tandem WW Domains in a Negative Regulator of Notch Signaling, Suppressor of Deltex. *J. Biol. Chem.*, 34991-35000.

[47] Patrie, K.M., Drescher, A.J., Welihinda, A., Mundel, P., and Margolis, B. (2002). Interaction of two actin-binding proteins, synaptopodin and alpha-actinin-4, with the tight junction protein MAGI-1. *J. Biol. Chem. 277*, 30183-30190.

[48] Craven, S.E., and Bredt, D.S. (1998). PDZ Proteins Organize Synaptic Signaling Pathways. *Cell 93*, 495-498.

[49] Fanning, A.S., and Anderson, J.M. (1999). PDZ Domains: Fundamental Building Blocks in the Organization of Protein Complexes at the Plasma Membrane. *J. Clin. Invest. 103*, 767-772.

[50] Garner, C.C., Nash, J., and Huganir, R.L. (2000). PDZ Domains in Synapse Assembly and Signaling. *Trends Cell Biol. 10*, 274-280.

[51] Doyle, D.A., Lee, A., Lewis, J., Kim, E., Sheng, M., and MacKinnon, R. (1996). Crystal Structures of a Complexed and Peptide-Free Membrane Protein–Binding Domain: Molecular Basis of Peptide Recognition by PDZ. *Cell 85*, 1067-1076.

[52] Tochio, H., Zhang, Q., Mandal, P., Li, M., and Zhang, M. (1999). Solution structure of the extended neuronal nitric oxide synthase PDZ domain complexed with an associated peptide. *Nat. Struct. Mol. Biol. 6*, 417-421.

[53] Hillier, B.J., Christopherson, K.S., Prehoda, K.E., Bredt, D.S., and Lim, W.A. (1999). Unexpected Modes of PDZ Domain Scaffolding Revealed by Structure of nNOS-Syntrophin Complex. *Science 284*, 812-815.

[54] Osten, P., Khatri, L., Perez, J.L., Köhr, G., Giese, G., Daly, C., Schulz, T.W., Wensky, A., Lee, L.M., and Ziff, E.B. (2000). Mutagenesis Reveals a Role for ABP/GRIP Binding to GluR2 in Synaptic Surface Accumulation of the AMPA Receptor. *Neuron 27*, 313-325.

[55] Harris, B.Z., and Lim, W.A. (2001). Mechanism and role of PDZ domains in signaling complex assembly. *J. Cell Sci. 114*, 3219-3231.

[56] Bixby, J.L. (2001). Ligands and signaling through receptor-type tyrosine phosphatases. *IUBMB Life 51*, 157-163.

[57] Glover, J.N., Williams, R.S., and Lee, M.S. (2004). Interactions between BRCT repeats and phosphoproteins: tangled up in two. *Trends Biochem Sci 29*, 579-585.

[58] Yuzawa, S., Ogura, K., Horiuchi, M., Suzuki, N.N., Fujioka, Y., Kataoka, M., Sumimoto, H., and Inagaki, F. (2004). Solution structure of the tandem Src homology 3 domains of p47phox in an autoinhibited form. *J. Biol. Chem. 279*, 29752-29760.

[59] Bruder, S., Linder, J.U., Martinez, S.E., Zheng, N., Beavo, J.A., and Schultz, J.E. (2005). The cyanobacterial tandem GAF domains from the cyaB2 adenylyl cyclase signal via both cAMP-binding sites9. *Proc Natl Acad Sci U.S.A. 102*, 3088-3092.

[60] Earles, C.A., Bai, J., Wang, P., and Chapman, E.R. (2001). The tandem C2 domains of synaptotagmin contain redundant Ca2+ binding sites that cooperate to engage t-SNAREs and trigger exocytosis. *J. Cell Biol. 154*, 1117-1123.

[61] Bedford, M.T., and Leder, P. (1999). The FF domain: a novel motif that often accompanies WW domains. *Trends Biochem Sci 24*, 264-265.

In: Focus on Cellular Signalling Research
Editor: Dorothy T. Leeds, pp. 131-153
ISBN 1-59454-619-3

Chapter VI

Calmodulin-Mediated Signaling in *Dictyostelium Discoideum*: CaMBOT Isolation and Characterization of the Novel Poly-Domain Protein Nucleomorphin and other Calmodulin Binding Proteins

Danton H. O'Day*
Department of Biology, University of Toronto at Mississauga
Mississauga, Ontario L5L 1C6, CANADA

Abstract

Calmodulin (CaM) is an essential protein in the model eukaryote *Dictyostelium discoideum* where it mediates numerous events including chemotaxis, gametogenesis, fertilization, and spore germination. Profiling using the CaM-binding overlay technique (CaMBOT) has revealed that well over four-dozen calcium-dependent and -independent CaM-binding proteins (CaMBPs) are present in *Dictyostelium*, some of which are linked to these CaM-mediated processes. Individual CaMBPs have been localized to specific sub-cellular locales and to show varying patterns of developmental expression. CaMBOT also is used to isolate cDNAs encoding CaMBPs from an expression library from mid-to-late multicellular development of *Dictyostelium*. Previously studied CaMBPs such as calcineurin A (CNA) and regulatory myosin light chain (RMLC) were identified in this way. Novel proteins were also identified. Nucleomorphin is a novel nuclear CaMBP, expressed as two main isoforms NumA1 and 2, which localizes to the nucleoplasmic periphery. NumA has at least one bipartite NLS (^{48}KKSYQDPEIIAHSRPRK^{66}) and a single CaM-binding domain (^{171}EDVSRFIKGKLLQKQQKIYKDLERF^{195}) that interacts with CaM in both a Ca^{2+}-dependent and Ca^{2+}-independent manner. NumA1 contains an

* Department of Biology, University of Toronto at Mississauga. Mississauga, Ontario L5L 1C6, CANADA. Email: doday@utm.utoronto.ca

extensive glu/asp or DEED repeat that regulates nuclear number. Yeast two hybrid and co-immunoprecipitation studies have identified calcium binding protein CBP4a as an interacting protein that binds to the DEED repeat. NumA2 also has a breast cancer C-terminus (BRCT) domain, two poly-N tracts plus a second, putative CaMBD. Proteins not previously known to be CaMBPs were also identified by CaMBOT probing of the *Dictyostelium* developmental expression library, including phosphoglycerate kinase, thymidine kinase and histone H1. *Dictyostelium* phosphoglycerate kinase (DdPGK) binds CaM in a Ca^{2+}-dependent manner with CaM negatively regulating its activity *in vitro*. The identified CaMBD shows 80% identity with PGKs from diverse organisms and is localized adjacent to mutation sites underlying certain human diseases. Thymidine kinase (DdTK1) is a Ca^{2+}-dependent CaMBP and its *in vitro* activity is slightly enhanced by CaM and inhibited by CaM antagonists. Histone H1 (DdH1) was isolated as a CaMBP and binds CaM *in vitro* but its mode of CaM-binding remains elusive. Each of the CaMBPs we have identified is discussed in terms of its individual roles as well as its potential interaction with other CaMBPs and other proteins in regulating specific CaM-dependent cellular processes including the cell cycle.

Key words: Nucleomorphin, calmodulin binding protein, phosphoglycerate kinase, histone H1, nuclear number, NLS, DEED repeat, calcium-binding protein, BRCT domain, WW interaction motif, cell cycle

Introduction

CaM-Mediated Signaling in *Dictyostelium*

Dictyostelium is a eukaryotic microbe that possesses a multitude of attributes making it an excellent organism for the study of cellular and developmental processes. The sequencing of the *Dictyostelium* genome has been reported further supporting this role as a model eukaryote for studies on the cell biology and development of higher metazoans (Williams et al, 2005). Recently, *Dictyostelium* was identified as a model organism by NIH indicating that it will provide valuable biomedical insights. While the specifics of calcium signaling remain to be resolved in *Dictyostelium*, the evidence supporting a central and essential role for CaM is redoubtable. CaM is an essential protein in this model eukaryotic microbe since gene disruption results in a lethal phenotype (Clarke, 1990). Compared to human CaM, *Dictyostelium* CaM shares 89% identity and, other than having three additional amino acids, it is over 99% similar and functionally identical. Thus studies on *Dictyostelium* CaM, like the organism itself, have relevance to understanding calmodulin function in higher organisms. In this chapter, we will focus on research from our laboratory to elucidate the current state of research on CaM and its target proteins in this model eukaryote. This work has progressed from the initial identification of CaM as a critical mediator of specific biological processes, through the profiling of the functional targets of CaM to the final identification of specific calmodulin binding proteins. While still a work in progress, some unique insights have been gained by this holistic biological approach.

Early work on CaM function in *Dictyostelium* was primarily concerned with the effects of antagonists on selected cellular and developmental events. Endogenous activation of

Dictyostelium spores via a diffusible autoactivator is CaM-dependent since it was inhibited by the classic CaM antagonists TFP and calmidazolium (Lydan and Cotter, 1995). During spore germination, the final emergence of amoebae is also CaM-dependent (Lydan et al, 1994b). CaM is a classic regulator of specific CaM-dependent kinases in many cellular processes but CaM-dependent phosphorylation does not appear to be involved in any aspect of spore germination (Lydan et al, 1994a). While the mechanism of CaM action during spore germination remains to be resolved a Ca^{2+}-dependent 64kDa CaMBP associated with the process of autoactivation has been identified (Lydan and Cotter, 1995). We have recently begun to characterize CaMBP64, a 64kDa protein, that is expressed during mid-to-late asexual development and which appears to be a cell surface protein (O'Day et al, unpublished; Table 1). Trifluoperazine and calmidazolium each dose-dependently inhibit cAMP and folic acid (FA) chemotaxis in *Dictyostelium* suggesting a role for CaM in chemoattractant-mediated cell movements (Gauthier and O'Day, 2001). These studies also implicated CaM-dependent phosphorylation as being central to cAMP-mediated chemotaxis.

Table 1. *Dictyostelium* calmodulin binding proteins isolated using the calmodulin binding overlay technique

CaMBP(Gene)	GenBank/ DictyBase #	Attributes/ Locale	Important Domains	Cell/Dev. Function
Calcineurin (*canA, pp2B*)	U22397/ DDB0001631	Protein phosphatase/ cytosol, nucleus	ser/thr protein phosphatase signature	growth? development?
Histone H1(*hstA*)	U50504/ DDB0191459	basic protein/nuclear	none	chromatin condensation, Transcript. reg.
Nucleomorphin (*NumA*) isoforms	AF140042/ DDB0231257	Novel/nuclear	DEED, NLS BRCT	Affects number of nuclei
CaMBP46 (*cmbB*)	AF140043/ DDB0201645	Novel/unknown	IP22 repeats LRR-like)	unknown
CaMBP16 (*cmbC*)	AF140044/ DDB0191521	Novel/unknown	none	unknown
Phosphoglycerate kinase (*pgkA*)	AAG34561/ DDB0191349	phospho-transferase cytosol/nucleus	ATP- binding	ATP prod., reg DNA synthesis
CaMBP64	AF318287/ DDB0214923	Novel/unknown	cys-rich, EGF-like	cell adhesion?
Thymidine kinase 1 (*thyB*)	AF510846_1/ DDB191436	kinase/cytosol	P-Loop	proliferation, apoptosis
Sugar transporter (*symA*)	AF510847/ DDB0215383	transporter/cell membrane?	Peptidase S8/S53	substrate transport?

These calmodulin binding protein were all isolated from the expression library of multicellular development from *Dictyostelium* as previously detailed (Myre and O'Day, 2002; O'Day, 2003). Searching DictyBase (http://dictybase.org) with the appropriate identity number will yield all of the current information and references that are available for each of the proteins listed in this table

While the majority of labs internationally study asexual development, *Dictyostelium* also has a sexual life style in which calcium signaling is central (O'Day et al, 1995). Agents that affect calcium mobilization prevent the developmental accumulation of glycoproteins linked to sexual development and CaM-dependent protein phosphorylation is associated with specific developmental events (Browning and O'Day, 1991; Lydan and O'Day, 1993b). As for many other biomembrane fusion events, calcium is essential for the fusion of gametes and for zygote formation (Lydan and O'Day, 1988a). In keeping with this pharmacological studies have shown that CaM is required for both gamete and pronuclear fusion (Lydan and O'Day, 1988b). In contrast, calcium is not required for gamete formation and CaM appears to be a negative regulator of gametogenesis since different CaM antagonists lead to the production of greatly inceased numbers of gametes (Lydan and O'Day, 1988b). An endogenous regulator of cell and pronuclear fusion that is secreted by the zygote is an antagonist of CaM function (Lydan and O'Day, 1989). To further support these central developmental roles of CaM, the endogenous autoinhibitor enhances gamete formation, a CaM-inhibited event, but inhibits the CaM-dependent events of cell and pronuclear fusion. Cellular and developmental studies using pharmacological approaches reveal the importance of CaM function but do not identify the specific effectors, the calmodulin binding proteins (CaMBPs), that do CaM's work.

Calmodulin Binding Proteins of *Dictyostelium*

The CaM-binding overlay technique (CaMBOT) is a useful tool for profiling essentially the complete complement of calcium-dependent and -independent CaMPBs in cells, tissues and organs after SDS-PAGE (O'Day, 2003). CaMBOT profiling has revealed that *Dictyostelium* contains over 4 dozen CaMBPs some of which are linked to specific cellular and developmental processes during asexual development, spore germination and sexual development (Lydan et al, 1994b; Lydan and O'Day, 1993a; Figure 1). Coupling pharmacological approaches and subcellular fractionation with CaMBOT has provided insight into the role of CaM and its CaMBPs during chemotaxis (Gauthier and O'Day, 2001). CaMBOT profiles of subcellular (cytosol, membrane, nuclear) fractions were compared between cells undergoing asexual multicellular development on a solid substratum with starved (able to undergo cAMP-mediated chemotaxis) and re-fed cells (able to undergo folic acid-mediated chemotaxis) to determine if certain CaMBPs were linked to either cAMP or folic acid chemotaxis. Of the over four dozen CaMBPs in cell extracts, most of the Ca^{2+}-dependent CaMBPs were cytosolic with four (43, 46, 66, and 78 kDa) linked to cAMP chemotaxis but none to FA chemotaxis. One 200 kDa nuclear Ca^{2+}-dependent CaMBP was associated with cAMP chemotaxis and one 34kDa Ca^{2+}-dependent CaMBP linked to FA chemotaxis. No Ca^{2+}-independent CaMBPs were linked to FA chemotaxis but three (27, 32, 40 kDa) cytosolic, a 16 kDa membrane and an 18 kDa nuclear Ca^{2+}-independent CaMBP were associated with cAMP-mediated chemotaxis. While these data link unique CaMBPs to events of chemotaxis, they don't identify what those CaMBPs are or their specific roles.

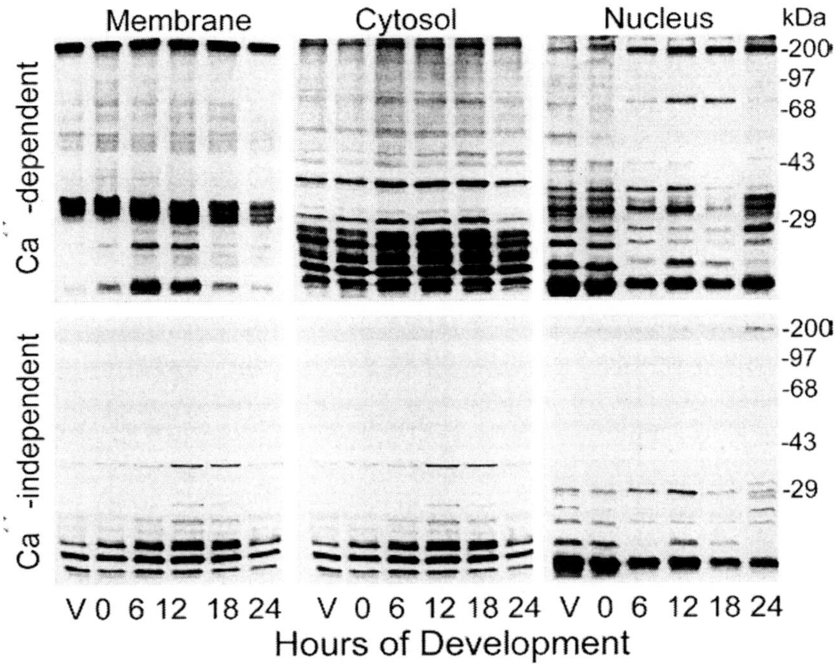

Figure 1. Calmodulin Binding Profiles during multicellular development of *Dictyostelium discoideum*. Cells were harvested from growing (V, vegetative) cultures, washed (0 hours) and plated on agar after which samples were removed every 6 hours. The samples were fractionated into membrane, cytosol and nuclear fractions that were run on SDS-PAGE and probed in the presence of calcium (Ca^{2+}-dependent CaMBPs) or EGTA (Ca^{2+}-independent CaMBPs) using the CaM-binding overlay technique (Gauthier and O'Day, 2001; O'Day, 2003). Positions of molecular weight markers are shown on the right

As introduced above, sexual development of *Dictyostelium* is a Ca^{2+}-dependent multicellular event involving gamete fusion, zygote differentiation, chemotaxis, phagocytosis and zygote maturation (O'Day et al, 1995; Szabo et al, 1987). CaMBOT profiling has revealed that dozens of Ca^{2+}-dependent CaMBPs also are expressed during sexual development many of which are developmentally regulated (Lydan and O'Day, 1993a). The expression of only seven of these CaMBPs was found to be regulated by the presence of calcium in the culture medium suggesting these few may be critical to the events of sexual development. Of the known CaMBPs, spectrin/fodrin was identified and implicated in gamete fusion. Of the unknown CaMBPs, one (CaMBP98) was present only during the acquisition of fusion competence, one (CaMBP48/46) was present only at the time of fusion and one (CaMBP38) during fusion through zygote giant cell differentiation. We are currently attempting to determine if CaMBP48/46 is phosphoglycerate kinase and if CaMBP38 is the small isoform of nucleomorphin, two proteins discussed below (Myre and O'Day, 2004c). Like asexual development, a group of rapidly migrating CaMBPs at molecular weights below 35kDa was identified in sexual cells but their expression levels appeared constant throughout development. One of these low molecular weight CaMBPs may be histone H1 which binds to CaM in *Dictyostelium* (O'Day and Myre, 2004; Table 1). While it is interesting to speculate on what CaMBPs are identified by CaMBOT profiling, their true identities require other approaches.

In spite of the importance of Ca^{2+} and CaM for growth and development in *Dictyostelium*, only a few CaMBPs have been characterized in any detail by other laboratories. Various studies have shown that cell motility and/or contraction in *Dictyostelium* cells are mediated by potential CaM-dependent effectors including actin binding protein fodrin (Bennett and Condeelis, 1988), IQGAP-related protein DGAP1 (Faix et al., 1998), plus the myosin I isoforms myoA-E (Coté et al., 1985) and myoJ (Hammer et al., 1996). Calcineurin A (CNA), or CaM-dependent protein phosphatase 2B, is one of the most well studied CaMBPs to date. CNA has been implicated in cell signaling events, asexual development and, more specifically, cAMP-mediated chemotaxis in *Dictyostelium* (Dammann et al, 1996; Polesky and O'Day, 1995; Weissenmayer et al, 2005). Other potential CaMBPs have been isolated but their binding to calmodulin has not been experimentally verified. One cannot assume that a protein shown to be a CaMBP in other organisms is necessarily a CaMBP in all organisms. For example, myosin light chain kinase, a well documented CaMBP in higher eukaryotes, does not bind to and is not regulated by CaM in *Dictyostelium* (Manalan and Klee, 1984; Tan and Spudich, 1990). Since at least four dozen Ca^{2+}-dependent and Ca^{2+}-independent CaMBPs have been shown to exist in *Dictyostelium* much more work remains to be done to identify and verify the true complement of CaMBPs in this organism and to determined their functional importance. One approach to doing this involves methods that lead to the selective identification of CaMBPs.

CaMBOT: Isolation of cDNAs Encoding CaMBPs

The use of CaMBOT probing of a cDNA expression library provides the ability to isolate and characterize cDNAs specficially encoding CaMBPs in *Dictyostelium* as well as in other organisms (Figure 2; Table 1; O'Day, 2003). This approach has been successfully used in identifying novel CaMBPs in a number of plant species (Zielinski, 1998; Arazi *et al*, 1999, 2000). Initial screening by our laboratory of a λZAP cDNA expression library from multicellular development of *Dictyostelium* using ^{35}S-radiolabeled recombinant CaM led to the isolation of a large number of cDNAs encoding presumptive CaMBPs. The first CaMBP to be isolated using this approach encoded the Ca^{2+}-CaM-dependent protein phosphatase calcineurin (CNA), a protein that is under active studies by others (Table 1; GenBank accession #U22397; Lydan and O'Day, 1994; Dammann et al, 1996). We have continued exploiting CaMBOT probing to isolate and characterize CaMBPs from *Dictyostelium* many of which are listed in Table 1. As we will see, CaM's target proteins in *Dictyostelium* often show structural and functional identity to CaMBPs of higher organisms and they mediate processes of central biomedical importance. Thus the study of CaMBPs in this model organism will provide insight into their function in all eukaryotes. With this in mind, this chapter will focus some of the CaMBPs that have been characterized in our lab as we continue to dissect the role of calmodulin in fundamental biological processes in *Dictyostelium*.

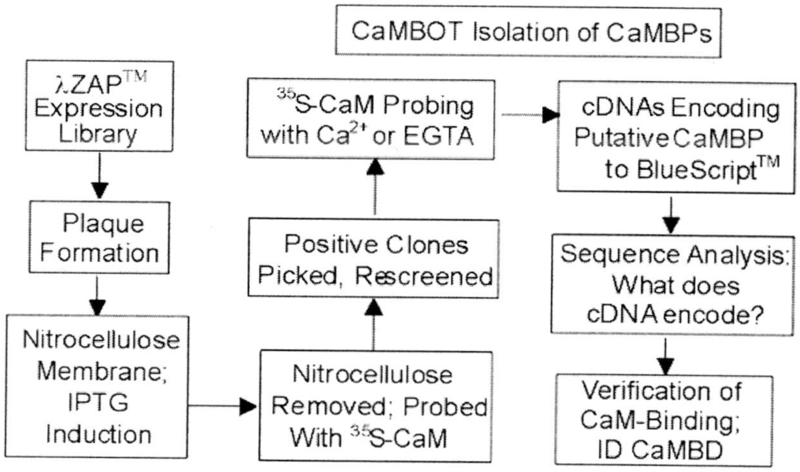

Figure 2. CaMBOT isolation of cDNAs encoding putative calmodulin binding proteins from expression libraries. The sequence of events in isolating cDNAs encoding putative CaMBPs and their verification are outlined (Myre and O'Day, 2002; O'Day, 2003)

Current Research

Nucleomorphin: A Novel Nuclear CaMBP

Nucleomorphin (NumA) represents the most fully characterized CaMBP of *Dictyostelium*. This acidic nuclear protein was isolated from the multicellular developmental expression library using CaMBOT (Myre and O'Day, 2002). Initially, nucleomorphin (originally NumA, now NumA1) was isolated as a cDNA (1119bp) encoding a 38.7kDa (340aa) protein. Subsequently, two other isoforms NumA2 and NumA3 (originally NumB and NumC, respectively) were found to be expressed differentially from NumA1 (Figure 3; Myre and O'Day, 2004c). Database searches revealed that nucleomorphin is a novel protein. Sequence analysis of NumA1 using the Calmodulin Target Database originally predicted three putative CaM-binding domains (CaMBDs) and the binding of NumA1 to CaM was verified experimentally (Myre and O'Day, 2002; Yap *et al*, 2000). To identify which region of NumA was responsible for the binding to CaM we generated a series of T7-tagged NumA deletion constructs that were expressed in *E. coli* and tested for CaM-binding (Myre and O'Day, 2005). We found that NumA was retained on CaM-agarose in both a Ca^{2+}-dependent and Ca^{2+}-independent manner. The amount of NumA retained on CaM-agarose was relatively the same in the presence of either Ca^{2+} or EGTA. The binding to CaM could be abolished by the addition of the potent CaM-inhibitor W-7. Further to this, constructs missing a region comprised of 47 residues between Leu^{182} to Phe^{195} were not retained on CaM-agarose in either the presence or absence of Ca^{2+}. This region encompasses a sequence (^{171}EDVSRFIKGKLLQKQQKIYKDLERF^{195}) with a distribution of hydrophobic amino acids (underlined) fitting a 1-16 motif appropriate for mediating Ca^{2+}-dependent CaM-binding (Fig. 3, CaMBD1). In addition this region also contains an overlapping sequence fitting a loose IQ motif (^{182}LQKQQKIYKDLERF^{195}) for Ca^{2+}-independent CaM-binding (Bahler and

Rhoads, 2002; Myre and O'Day, 2004). While the role of CaM-binding has not yet been resolved for nucleomorphin, previous studies had shown that, as discussed below, the CaMBD of certain nuclear CaMBPs can function in nuclear translocation (Agentaro et al, 2003).

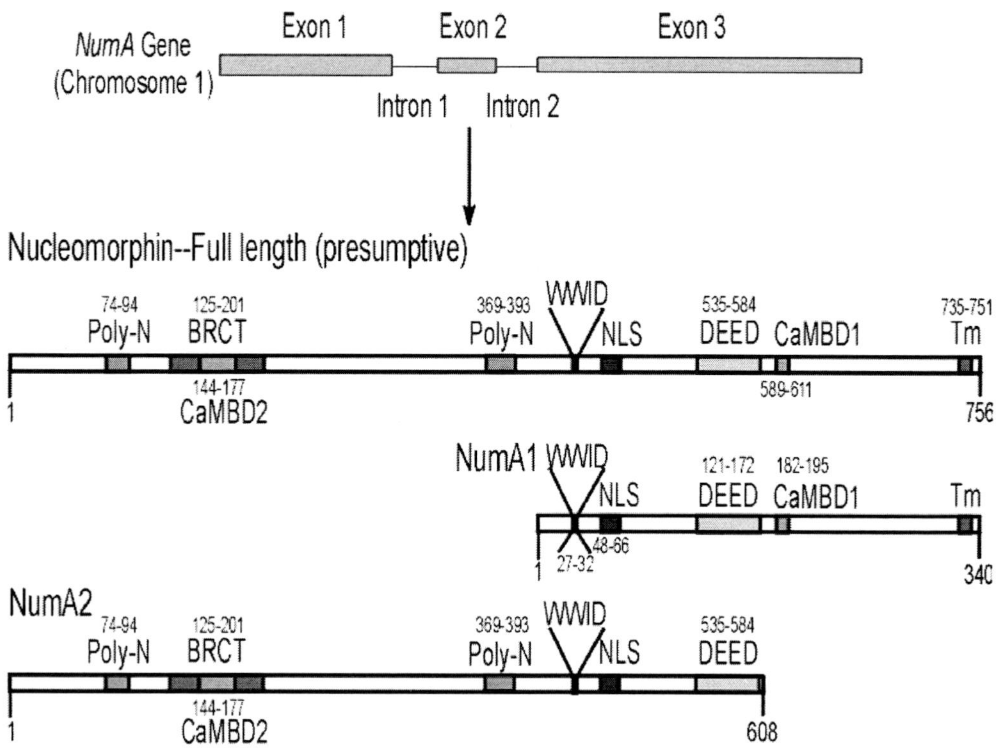

Figure 3. The nucleomorphin gene (NumA) and its isoforms. The nucleomorphin gene is comprised of three exons and two introns. The full presumptive full length nucleomorphin protein has not been shown to be expressed but, if it were, it would contain all of the domains found in both NumA1 and NumA2. Previously identified domains shown are the breast cancer carboxy-terminus (BRCT) domain, the nuclear localization sequence (NLS), glu/Asp (DEED) repeat, putative transmembrane domain and the first CaM-binding domain (CaMBD1). Previously un-delineated domains include the poly-asparagine (Poly-N), CaMBD2 and the class IV WW interaction motif (WWID)

Nuclear Localization of Nucleomorphin Requires an NLS

In many cases, the active import of a protein into the nucleus requires the presence of a nuclear localization sequence (NLS) but previously no NLS had been experimentally identified for any protein of *Dictyostelium*. While several nuclear proteins, including Dd-STAT proteins, skp1 and Hsp32 have been identified and functionally characterized in *Dictyostelium*; none have been shown to possess an experimentally determined NLS (Zhukovskaya *et al*, 2003; Van Der Wel *et al*, 2002; Moerman and Klein, 1998). Analyses of NumA using PSORT and PSORTII suggested the presence of four possible nuclear localization signals of which two were identified by both programs (Myre and O'Day, 2002; Horton and Nakai, 1997). To determine if any of were true nuclear targeting signals and to assess the role of the CaMBD in the nuclear import of nucleomorphin, a series of NumA1 deletion constructs fused to the green fluorescent protein (GFP) were generated. N-terminal residues 1-129 were sufficient to localize GFP-constructs to the internal periphery of nuclei as observed with full length GFP-NumA1 constructs (Figure 4A, D, E; Myre and O'Day, 2002, 2005). Thus the identified CaMBD is not directly involved in the nuclear translocation or subnuclear localization of NumA1. The N-terminal region contains the amino acid residues ^{48}KKSYQDPEIIAHSRPRK^{64} that include both a putative bipartite and classical NLS (Figure 3, NLS; Myre and O'Day, 2005). When this sequence alone was coupled to GFP it led to the intranuclear localization of GFP however it did not lead to its subcellular localization as patches adjacent to the inner nuclear envelope (Figure 4C). Deletion of the N-terminus containing this sequence prevents the movment of NumA constructs into the nucleus (Figure 4B). Thus, while the experimentally verified NLS determined by that work is sufficient to localize NumA to the nucleus, it does not reveal how NumA becomes localized to the internal periphery of nuclei. Truncated forms of GFP-NumA lacking the putative transmembrane domain also localized to the nuclear periphery revealing that it is not directly involved in the event. As discussed below (Localization of NumA1 to the Inner Nuclear Membrane: An Hypothesis) a putative WW domain interaction could be involved in the subnuclear localization of NumA (Figure 3, WWID).

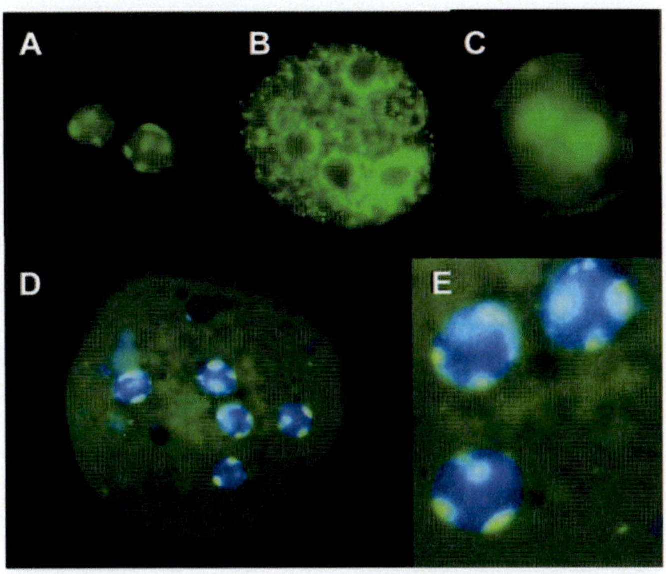

Figure 4. The intracellular localization of GFP-constructs of nucleomorphin. A. Full length GFP-NumA constructs localize within the nucleus predominantly as bright patches adjacent to the nuclear envelope. B. GFP-NumA1 constructs lacking the NLS (GFP-ΔN160) localize to the cytoplasm. The dark circles surrounded by fluorescence appear to be vesicles since they do not fluoresce with the nuclear stain Hoechst 33258. C. GFP-coupled to an NLS localizes within the nucleus but not at the nuclear periphery. Two adjacent nuclei are visible in this image and the limited amount of cytoplasmic fluorescence may be artifact caused by the bright nuclear fluorescence. D. Localization of GFP-Δ118-167 in Hoechst-stained (blue) nuclei. E. Closeup of nuclei showing that removal of the DEED repeat alone does not alter the subnuclear localization of the fluorescence

As yet, the potential role of other NLS or mechanisms that might be involved in the nucleocytoplasmic transport have not been ruled out and the presence of export signals has not been addressed. We are currently testing the ability of the NLS to translocate GFP-linked peptides into the nuclei of various eukaryotic cells.

Nucleomorphin Works within the Nucleus to Regulate Nuclear Number

The analysis of the various constructs has led to the discovery of one function of NumA1: it regulates the number of nuclei that are produced by cells (Myre and O'Day, 2002). We are also gaining insight into how it is able to do this. A priori, the localization of nucleomorphin as distinct patches at the nuclear periphery and the presence of a defined NLS that directs intranuclear delivery suggest that NumA has a nuclear function. Since extensive attempts to knockout NumA1 were unsuccessful, various deletion constructs were made in an attempt to gain insight into its function. The extensive glu/asp (DEED) repeat of NumA1 is particularly interesting because of its structure and because this repeat bears similarity to that seen in other nuclear proteins including nucleoplasmin and nucleophosmin (Figure 3, DEED; Myre and O'Day, 2002). Nucleoplasmin and nucleophosmin serve as nucleohistone chaperones transferring histones to nuclear DNA forming nucleohistones. For example, nucleoplasmin is an acidic, nuclear protein that has been shown to mediate chromatin condensation as well as the replacement of somatic histone H1 by an oocyte-specific linker histone (e.g., Dimitrov and Wolffe, 1995). An examination of NumA constructs lacking and possessing various domains revealed that it is involved in regulating the number of nuclei produced by cells.

Liquid cultured axenic cells of *Dictyostelium* are over 70% mononucleate with around 25% binucleates with the remainder being multinucleates primarily with 3-4 nuclei. The presence of binucleate cells in axenic strains of *Dictyostelium* grown in liquid culture may be due to the absence of a solid substratum upon which to effect cytokinesis. Full length GFP-NumA constructs generate equivalent numbers of nuclei to that seen in normal axenic cells (Figure 5, NumA). Removal of the C-term NLS likewise does not alter the number of binucleates versus multinucleates (Figure 5, ΔC36). However while the removal of the DEED repeat alone does not alter the subnuclear localization of the fluorescence it does result in multinuclearity with approximately 95% of the cells being multinucleate (Figure 5, Δ118-167; Myre and O'Day, 2002). GFP-constructs which possess the N-terminus containing the bipartite NLS but not the CaMBD or the DEED repeat localize to the nuclear periphery and express a multinucleate phenotype with as many as 16 nuclei present in a single cell (Figure 5, ΔC211; Myre and O'Day, 2002, 2005). On the other hand, cells expressing a construct lacking both the DEED domain and the amino-terminal NLS were like parental cells with mainly (~70.3±6%) uninucleate and binucleate (~24.4±3.1%) cells and very few (~2.3%) multinucleates (Figure 5, ΔN160; Myre and O'Day, 2005). Thus the removal of the DEED domain alone is not sufficient to induce multinuclearity, the DEED-less NumA construct must enter the nucleus to be effective. In other words, whatever the role of NumA is in regulating nuclear number, it does so through its intranuclear interactions.

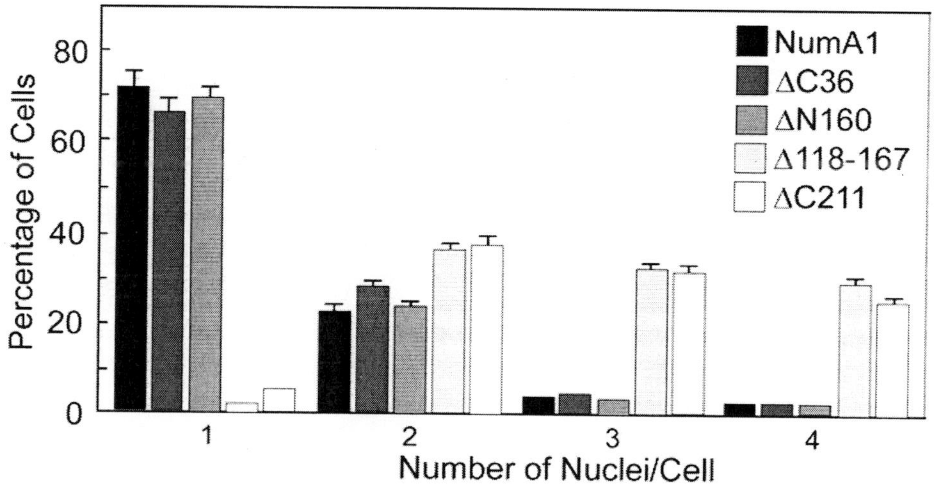

Figure 5. The number of nuclei found in cells expressing different nucleomorphin constructs. Cells containing the constructs indicated on the graph were grown in axenic medium and the number of nuclei per cell counted during the log phase of growth. Error bars are shown only when they extend beyond the bars

Multinucleate mutant phenotypes in *Dictyostelium* have been generated by others but these are linked to cytoplasmic events most notably changes in cytoskeletal elements that are essential for cytokinesis. Reported strains known to produce multinucleates in vegetatively growing cells include myosin II null or antisense mutants, and DdRasGAP1 mutants (de Lozanne, 1987; Knecht and Loomis, 1987; Larochelle *et al*, 1996; Lee *et al*, 1997). Similarly both a human cell line expressing a constitutively activated mutant of CDC42Hs and a *Dictyostelium* strain lacking the *Race* gene encoding a Rac/CDC42-related protein produce giant and multinucleate cells by impairing cytokinesis (Dutartre *et al.*, 1996; Larochelle *et al.*, 1996). The involvement of NumA in regulating nuclear number does not appear to be due to alterations in cytokinesis because construct cells expressing the multinucleate phenotype grow and divide in suspension at a rate similar to parental AX3 cells. This and experimental evidence showing that these constructs must enter the nucleus to affect multinuclearity imply that the multinucleate phenotype caused by deletion of the DEED domain is the result of nuclear not cytoskeletal events. Further insight into the function of the DEED repeat has been gained through the use of yeast two-hybrid screening and coimmunoprecipitation studies.

Numa Interacts with CBP4a via the DEED Repeat

To further investigate the function of nucleomorphin and the DEED repeat, we screened a *Dictyostelium* cDNA library to identify NumA-interacting proteins using yeast two-hybrid analysis (Fields and Song, 1989). Using full length NumA1 as bait, a complete cDNA was isolated that encoded calcium binding protein 4a (CBP4a; Myre and O'Day, 2004b). While *Dictyostelium* possesses at least thirteen calcium-binding EF-hand proteins, with the exception of CaM, little is known of their cellular or developmental functions (Coukell *et al*, 1995; Andre *et al*, 1996; Sakamoto *et al*, 2003). CBP4a lacks a traditional NLS but since it is a small (18.7 kDa) hydrophilic protein it likely is able diffuse into the nucleus via nuclear pores. Subsequent two-hybrid analysis showed that NumA constructs lacking the DEED domain failed to bind to CBP4a (Myre and O'Day, 2004b). The DEED-mediated interaction between CPB4a and NumA was also demonstrated *in vitro* using various T7-tagged NumA deletion constructs (Myre and O'Day, 2004b). When the *in vitro* pull-down assays were carried out in the presence of EGTA, CBP4a and nucleomorphin did not coimmunoprecipitate revealing that their interaction is Ca^{2+}-dependent (Myre and O'Day, 2004b). This is the first evidence of a potential role for CBP4a as a mediator of nuclear number with the interaction between CBP4a and NumA being mediated by intranuclear Ca^{2+}/CaM-mediated signal transduction.

Localization of NumA1 to the Inner Nuclear Membrane: An Hypothesis

One candidate that may determinine the intranuclear localization of NumA1 to the nuclear envelope is a class IV WW domain interaction motif (^{27}EGTTKPK^{32}) that exists in its N-terminus (Figure 3, WWID; O'Day, unpublished). As its name implies this motif is linked to protein-protein interactions mediated by class IV WW domains (Sudol et al, 2001). Class IV WW domains bind to such sequences when the proline is preceded by phospho-S or phospho-T and, in keeping with this, the identified class IV WW domain interaction motif of NumA1 encompasses GSK3 (^{26}DEGTTKPKS^{33}) and proline-directed/MAPK (^{27}EGTTKP KS^{33}) phosphorylation sites. Nuclear WW-domain containing proteins also have been shown to be involved in the regulation of cell cycle events (Sudol et al, 2001).

As mentioned, yeast two hybrid analysis of NumA-binding proteins led to the identification of CBP4a, a calcium binding protein that interacts with the DEED domain of NumA1 (Myre and O'Day, 2004b). Using the same approach, a cDNA encoding *Dictyostelium* puromycin-sensitive aminopeptidase (DdPsa; DictyBase # DDB0231238) was also found to be a NumA interacting protein (Myre and O'Day, unpublished). Visual scanning of the DdPsa shows a putative WW domain (^{332}WLNEGFATFMGYLVTDY LYPKW^{353}) suggesting that DdPsa may interact with NumA1 via this domain. DdPsa shares 65% sequence identity with DdPsb, a second puromycin-sensitive aminopeptidase (DictyBase # DDB0231240) that possesses an equivalent WW domain (^{327}WLNEGFATFMGYLSTHHLFPEW^{348}) with 77% identity to that of the WW domain of Psa. While their primary function is protein digestion, aminopeptidases have been implicated in a diversity of regulatory processes including protein processing, signal transduction, cancer cell proliferation and antigen presentation (Saveanu et al, 2002). For example, in the mouse *Psa* activity has been suggested to participate in proteolytic events essential for cell growth and viability (Constam et al. 1995; Sekine et al. 1999). Whether the Class IV WW domain interaction motif plays a functional part in NumA remains to be resolved.

Identification of a BRCT-Domain in Nucleomorphin Isoform NumA2

In attempt to understand more about the structure and function of nucleomorphin, the *Dictyostelium* database (DictyBase) was searched using the complete nucleomorphin nucleotide sequence reported by Myre and O'Day (2002; http://dictybase.org/db/cgi-bin/blast.pl). The gene encoding nucleomorphin resides on chromosome one with the coordinates 2485527 to 2488002. This region encompasses an additional 1443 base pairs upstream of the originally identified cDNA encoding the nucleomorphin isoform NumA1 (Figure 3; Myre and O'Day, 2002, 2004c). *NumA* exists as a single copy gene comprised of three exons interrupted by two short introns. In keeping with this, the sequence predicts a much larger protein composed of 756 amino acids with a molecular weight of ~85.4 kDa which was originally called NumB but renamed NumA2 in keeping with DictyBase gene and protein naming protocols (Myre and O'Day, 2004c). The structural relationship between NumA1 and 2 is detailed in Figure 3 revealing a few new regions not covered in previous work that are under analysis (e.g., Poly-N, Class IV WW domain interaction motif, and CaMBD2 in the BRCT domain). The nucleotide sequence of NumA2 encoding the C-terminal end (amino acids 417 – 756) overlaps with the N-term sequence of the NumA1 isoform including the experimentally determined DEED repeat and NLS (Figure 3). While NumA2 terminates just after the DEED repeat and does not seem to share the NumA1 CaMBD, a search using the Calmodulin Target Database and subsequent visual analysis, detected a potential CaMBD within the N-term of NumA2 (Figure 3, CaMBD2; O'Day, unpublished). This domain has not yet been experimentally validated. Sequence analysis identified a putative conserved breast cancer carboxy-terminus (BRCT) domain (Figure 3, BRCT; residues 126-203) in the amino-terminus. The NumA2 BRCT domain is highly homologous to known BRCT-domains of the human breast cancer susceptibility protein BRCA1 (Bork *et al*, 1997; Koonin *et al*, 1996). The basic biological function of the BRCT domains is still unknown. A DIG-labeled probe made to the BRCT-NumA-amino terminal domain identified both NumA2 and the NumA1 (Myre and O'Day, 2004c). Nucleomorphin anti-sera also recognized both NumA1 and A2 verifying that both transcripts are translated into functional proteins. The presence of a nucleomorphin isoform possessing the BRCT domain reinforces the potential role of this CaMBP in cell cycle events. If nucleomorphin does have a cell cycle function, then it should be differentially expressed during phases of growth, starvation and development. Recent research reveals that NumA isoforms do show independent patterns of expression that support this role.

NumA Expression in Response to Starvation and Refeeding

When nutrients are depleted, *Dictyostelium* cells placed on a solid substratum initiate a multicellular developmental program to ensure survival of the organism (Maeda, 1997). When cells are starved in a liquid medium they cannot undergo multicellular development but undergo a series of events that prepare them for multicellular development while simultaneously enhancing their chances of survival. In order to assess the effects of starvation on NumA1 and NumA2, gene-specific primers were designed to selectively amplify each isoform from total RNA using reverse transcriptase-PCR. When axenically grown cells are starved, the level of NumA1 expression is significantly decreased (~ 88%) within 4 hours. Upon refeeding, NumA1 levels quickly return to vegetative levels (Myre and O'Day, 2004c). In contrast when cells are starved on a solid surface that will allow the progression of multicellular development, NumA1 RNA expression does not decrease over 4 hours. Thus starvation can reduce NumA gene expression but not if cells are allowed to begin multicellular development. On the other hand, western blotting has shown that the level of NumA1 protein remains unchanged in response to starvation but when cells are refed bacteria after 4-hour of starvation two extra bands appear (50 and 38 kDa; Myre and O'Day, 2004c). These proteins possibly represent posttranslationally modified forms of NumA1 or are the result of proteolytic processing. Proteolytic cleavage and subsequent protein degradation are irreversible regulatory events reserved for key cellular processes including the cell cycle, response to nutritional change, some signaling pathways and apoptosis, whereas the effects of posttranslational modifications are typically reversible (Pickart, 2001). In contrast to NumA1, northern analysis shows that NumA2 is expressed minimally during vegetative growth and increases after 2 hours of starvation in buffer to reach a steady level. This is the first differentially expressed BRCT-domain containing protein in *Dictyostelium.*

To add to the story, a number of other CaM-binding proteins were also isolated from the same developmental expression library of *Dictyostelium*. While it is too early to ascribe collaborative functions related to their presence during the same developmental events, the identification of these known proteins has special implications since they were not previously known to bind to or be regulated by calmodulin. These proteins include phosphoglycerate kinase (PGK), thymidine kinase 1 (TK1) and histone H1, each of which has been linked to the cell cycle along with other cellular events.

Phosphoglycerate Kinase as CaMBP

The probing of the developmental expression library from *Dictyostelium discoideum* with radiolabelled CaM led to the isolation of a cDNA encoding a putative CaMBP with 68% sequence similarity (78% homology) to human PGK (Table 1; Myre and O'Day, 2004a). We thus named it DdPGK (EC 2.7.2.3; *pgkA*). DdPGK is a ubiquitous highly conserved, monomeric enzyme that has been isolated from a large diversity of species (Watson and Littlechild, 1990). In keeping with this, DdPGK possesses all of the significant, defining PGK domains including the hinge (FAKALE, 100% identity) and nucleotide binding region (69% identity). However PGK had not previously been identified as a CaMBP in any organism. PGK is an enzyme with several functions (see review: Kim and Dang, 2005). As an ATP-generating glycolytic enzyme, PGK contributes to the glycolytic and gluconeogenic pathways. PGK also has been identified as a component of the transcriptional machinery for mRNA synthesis and it also influences DNA replication and repair in the mammalian nucleus. Because of its multifaceted functions, understanding the regulation of PGK is important. Thus we needed to verify the nature and significance of its CaM-binding ability.

Scanning of the protein sequence of DdPGK using the Calmodulin Target Database led to the identification of a single, strong putative CaM-binding domain

(^{209}KPFLAILGGAKVSDKIKLIE^{228}). Full-length and deletion T7-epitope tag constructs of DdPGK were made to determine the true CaMBD and the bacterial expression products were tested for their ability to bind to CaM-agarose. All constructs containing the putative CaMBD bound to CaM-agarose in the presence of calcium but not in the presence of EGTA. In contrast, those constructs lacking the CaMBD failed to bind to CaM-agarose under either situation. Thus DdPGK is a Ca^{2+}-dependent CaMBP. Sequence comparisons revealed that the identified CaMBD of DdPGK shared 80% identity with the same region in a diversity organisms, suggesting CaM-binding could be a common trait. When the CaM-binding ability of rabbit and yeast PGKs were tested, each bound to CaM-agarose in a Ca^{2+}-dependent manner. Since the experimentally identified CaMBD is situated adjacent to the the hinge and within the ATP binding domain, the CaM-binding attribute appeared to be significant with the potential to have a regulatory function in PGK activity. In keeping with this, CaM inhibited yeast PGK activity *in vitro* while the antagonist W-7 abrogated this inhibition. While *in vivo* analyses remain to be done, these data place PGK in the realm of enzymes regulated by Ca^{2+}/CaM-mediated signaling. In addition, they have biomedical implications since PGK mutations have been linked to hemolytic anemia, central nervous system dysfunction and/or myopathy with many of the mutations underlying these diseases occurring close to the identified CaMBD (Fujii, 2000; Myre and O'Day, 2004a; Yoshida, 1975, 1996).

Thymidine Kinase as CaMBP

A cDNA encoding another kinase also was isolated using CaMBOT suggesting it too was a CaMBP. The cDNA contained an open reading frame of 951bp encoding a 227aa polypeptide (25.5 kDa; O'Day et al, 2005a). Sequence comparisons led to highly significant matches with cytosolic thymidine kinases (TK1; EC 2.7.1.21) from a diverse number of species. The most significant hits were for mammalian, including human TK1 (7e-56; 59% Identities; 75% Positives, 2% Gaps). Thus the encoded protein is *D. discoideum* thymidine kinase 1 (Table 1; DdTK1; gene *thyB*; O'Day et al, 2005a). TK1 is a cytosolic enzyme that phosphorylates thymidine to produce thymidine-5'-phosphate, a precursor in the synthesis of genomic DNA. DdTK1 has highly conserved thymidine-binding domain plus an ATP-binding P-loop that shows 100% sequence identity to human TK1. TK1 expression serves as a prognostic marker for various cancers because it is almost undetectable in resting cells but its level increases to peak during the S phase of proliferating cells (Bello, 1974; Hengstschläger et al, 1994). TK also phosphorylates many anti-neoplastic and anti-viral agents including bromodeoxyuridine, fluoro-deoxyuridine and azido-thymidine allowing them to be incorporated during DNA synthesis (Vazquez-Padua et al, 1989). Thus, among other things, TK1 is of great biomedical import because of these roles in cell proliferation and antiviral drug activation.

As discussed for other proteins above, based on the method of isolation using CaMBOT, DdTK1 was a CaMBP. To verify this, the amino acid sequence of DdTK1 was scanned for the presence of potential CaM-binding domains and a putative CaMBD (^{20}GKTTELIRRIKRFNFANKKC^{30}) was identified. Using T7-tag constructs as detailed above for PGK CaM-binding studies were carried out. Wild type DdTK1 plus two constructs

(DdTKΔC36, DdTKΔC75) possessing the putative CaMBD bound to CaM-agarose *in vitro* in the presence of Ca^{2+} but not EGTA revealing DdTK1 to be a Ca^{2+}-dependent CaMBP. Constructs lacking the CaM-binding region did bind to CaM-agarose. Analysis of a large number of species showed that the identified CaMBD is highly conserved across species overlapping the highly conserved ATP-binding (P-loop) domain (O'Day et al, 2005). Since the recombinant DdTK1 enzyme is active in vitro, is stimulated ~15% by CaM and slightly inhibited by CaM antagonists, the fact that the identified CaMBD overlaps the ATP-binding (P-loop) domain supports the idea that CaM may function to regulate TK activity in vivo in a diversity of species. The isolation of DdTK1 as a CaMBP from the same developmental phase as other cell-cycle related CaMBPs suggests also that they may interplay in the CaM-mediated signaling events that control the cell cycle in eukaryotes.

The Unexpected Relationship between H1 and CaM

As mentioned above, nucleoplasmin and nucleophosmin share similarities to nucleomorphin especially in the acidic DEED domain. Nucleoplasmin is a chaperone that mediates chromatin condensation and directs the replacement of somatic histone H1 by an oocyte-specific linker histone (Dimitrov and Wolffe, 1995). It may be relevant that probing of the same developmental library that led to the isolation of nucleomorphin, DdTK1, and DdPGK also led to the isolation of a cDNA encoding histone H1. While the former have been discussed, the isolation of histone H1 as a potential CaMBP was especially intriguing not only for its link to chromosome condensation and gene regulation but also as a potential cell cycle component regulated by calmodulin. Histone H1 belongs to the linker histone family that binds both the linker DNA between nucleosomes and adjacent nucleosomes in nuclear chromatin (Zlatanova *et al.*, 2000). Through this higher order stabilization, H1 has been suggested to have a general repressive effect on transcription but H1 can also regulate the activity of specific genes (Gao et al, 1998; Thomas, 1998). In addition to these roles in chromatin structure and gene regulation, recent research has shown that H1 also plays roles in DNA repair and synthesis (Siddhartha et al, 2002; Yamanaka et al, 2002). Various studies also have shown that histone H1 binds to certain proteins including Epstein-Barr nuclear antigen 2A (Grasser et al, 1993) and the tumor repressor lysyl oxidase (Giampuzzi et al, 2003).

The inter-relationship between H1 and CaM and the actual binding of the nucleoprotein to CaM has been suggested by previous work. Histone H1 inhibits CaM-dependent autophosphorylation and activation of CaMKII as well as the protein phosphatase activity of calcineurin in the nucleus (Rasmussen and Garen, 1993). Natsukari *et al.* (1995) showed that CaM and histones co-localize in the nuclei in rat PC12 cells. They also revealed increased phosphorylation of histone H1 that correlated with increased CaM levels as the cell cycle progressed from G1 to M phase. In spite of this the direct binding of CaM to histone H1 has not been previously demonstrated. Subsequent binding analyses with CaM-agarose revealed that DdH1 bound in both the presence of either Ca^{2+} or EGTA. Initial scans of the DdH1 sequence in the Calmodulin Target Database and by visual analyses revealed the presence of a strong and a weak CaMBD. However, deletion constructs lacking either or both CaMBDs still bound strongly to CaM-agarose (O'Day and Myre, 2004). While this suggests that neither of the putative CaMBDs is involved in CaM binding, there are other interpretations. First, in addition to continuous CaMBDs, novel forms of CaM-binding involving discontinuous CaMBDs have recently been discovered (Hoeflich and Ikura, 2003). Similarly, certain basic helix-loop-helix (bHLH) transcription factors have been shown to interact with CaM in an unusual fashion (Onions et al, 2000). This interaction involves binding of two CaM molecules to one bHLH dimer. Onions et al (2000) propose that one domain of CaM preferentially binds the basic sequence while the other binds the modulatory sequence of bHLH.

While the precise nature of the interaction between DdH1 and CaM remains to be clarified, it does not appear to be a non-specific event. What functions could this interaction serve? The most direct role would be for CaM to either positively or negatively regulate H1-dependent events such as chromatin organization and gene transcription. Alternatively, free H1 released from chromatin during the cell cycle could be involved in sequestering CaM, making it unavailable for interaction with other nuclear CaMBPs such as nucleomorphin and CNA, the latter of which has been linked to the dephosphorylation of specific transcription factors. Clearly the interaction of CaM and H1 could have important implications to calcium/CaM-mediated signaling during the cell cycle in all organisms and their relationship within the nucleus needs to be resolved.

Conclusion

Research on Ca^{2+}/CaM-mediated signal transduction in *Dictyostelium* is progressing well. The ability to study a multitude of fundamental biological processes in such a well characterized, widely accepted and easily manipulated model organism is particularly useful. While profiling CaMBP populations during alternative cellular and developmental events indicates candidates worthy of further study, the identification and verification of proteins as true CaMBPs is a critical first step in understanding the role of this subset of signaling proteins in cellular and developmental functions. To date, only a handful of proteins have been verified as true CaMBPs in *Dictyostelium* and fewer CaM-binding domains have been experimentally determined. Nucleomorphin stands out as the best-characterized CaMBP and represents the first protein to have an experimentally identified NLS in this organism. Ongoing work has shown that NumA isoforms are differentially regulated by developmental morphogens (e.g, cAMP, DIF-1 and ammonia) opening the door for further understanding of the signaling mechanisms that oversee the function of this Ca/CaM-regulated protein (Myre, 2005). The results to date also reveal that, in general, research on calmodulin and its binding proteins in *Dictyostelium* will have biomedical value. Since several of the CaMBPs of *Dictyostelium* are either highly similar to their mammalian counterparts and/or share CaM-binding domains that are highly conserved across species, studies of those proteins will provide valuable insight into their function in higher cells. Finally, the further characterization of the BRCT-domain protein nucleomorphin and its interactions with its binding proteins and co-regulation with other CaMBPs should provide further insight into nuclear functions during the cell cycle and other biomedically relevant processes.

Acknowlegements

I would like especially like to thank Mike Myre for his contribution to many aspects of this work. I'd also like to acknowledge the contribution of Elizabeth White, Carla Loftus, Raymond Lim and Christopher Henley for their past assistance in parts of the work reported here. The research was supported by grants from the Natural Sciences and Engineering Research Council of Canada.

References

Andre, B., A.A. Noegel and M. Schleicher, 1996; *Dictyostelium discoideum* contains a family of calmodulin-related EF-hand proteins that are developmentally related, *FEBS Letters* 382: 198-202.

T. Arazi, R. Sunkar, B. Kaplan, H. Fromm, 1999. A tobacco plasma membrane calmodulin-binding transporter confers Ni^{2+} tolerance and Pb^{2+} hypersensitivity in transgenic plants. *Plant J.* 20: 171-82.

Arazi, T., B. Kaplan and H. Fromm, 2000. A high-affinity calmodulin-binding site in a tobacco plasma-membrane channel protein coincides with a characteristic element of cyclic nucleotide-binding domains. *Plant Mol Biol.* 42: 591-601.

Bahler, M., and A. Rhoads, 2002. Calmodulin signalling via the IQ-motif, *FEBS Letters* 513: 107-113.

Bello, L. J., 1974. Regulation of thymidine kinase synthesis in human cells. *Exp. Cell Res.* 89: 263-274.

Bennett, H., and J. Condeelis, 1988. Isolation of an immunoreactive analogue of brain fodrin that is associated with the cell cortex of *Dictyostelium discoideum*, *Cell Motil. Cytoskel.* 11: 303-317.

Browning, D.D., and D.H. O'Day, 1991. Con A and WGA binding glycoproteins associated with cell fusion and zygote differentiation in *Dictyostelium discoideum*: Effects of calcium ions and tunicamycin on glycoprotein profiles. *Biochem. Cell Biol.* 69: 202-290.

Bork, P., K. Hofmann, P. Bucher, A.F. Neuwald, S.F., Altschul, and E.V. Koonin, 1997. A superfamily of conserved domains in DNA damage-responsive cell cycle checkpoint proteins. *FASEB J.* 11: 68 - 76.

M. Clarke, 1990. Calmodulin structure, localization and expression in *Dictyostelium discoideum*, pp. 375-391 In: D.H. O'Day (ed.), *Calcium as an Intracellular Messenger in Eukaryotic Microbes,* ASM Press, Washington, D.C.

Constam, D.B., A.R.Tobler, A. Rensing-Ehl, I. Kemler, L.B. Hersh and A. Fontana, 1995. Puromycin-sensitive aminopeptidase. Sequence analysis, expression, and functional characterization. *J Biol Chem* 270: 26931–26939.

Côté, G.P., J. P. Abensi, T. Ueno, J. A. Hammer and E. D. Korn, 1985. Purification from *Dictyostelium discoideum* of a low-molecular weight myosin that resembles myosin I from *Acanthamoeba castellanii*, *J. Biol. Chem.* 260: 4543-4546.

Coukell, B., J. Moniakis and A. Grinberg, 1995. Cloning and expression in *Escherichia coli* of a cDNA encoding a developmentally regulated Ca^{2+}-binding protein from *Dictyostelium discoideum, FEBS Letters* 362: 342-346.

Dammann, H., S. Hellstern, Q. Husain and R. Mutzel, 1996. Primary structure, expression and developmental regulation of a *Dictyostelium* calcineurin A homologue. *Eur. J. Biochem.* 238: 391-399.

Dimitrov, S., and Wolffe, A.P., 1995. Chromatin and nuclear assembly: Experimental approaches towards there constitution of transcriptionally active and silent states. *Biochim. Biophys. Acta* 1260: 1–13.

Dutartre, H., J. Davoust, J. P. Gorvel and P. Chavrier, 1996. Cytokinesis arrest and redistribution of actin-cytoskeleton regulatory components in cells expressing the Rho GTPase CDC42Hs. *J. Cell Sci.* 109: 367 – 377.

Fields, S., and O. Song, 1989. A novel genetic system to detect protein-protein interactions, *Nature* 340: 245-246.

Fujii, H., 2000. Other erythrocyte enzyme deficiencies associated with non-haematological symptoms: phosphoglycerate kinase and phosphofructokinase deficiency, Bailliere's *Clin. Haematol.* 13: 141-148.

Gao, B., H. Jaffe and G. Kunos, 1998. Histone H1 isoforms purified from rat liver bind nonspecifically to the nuclear factor 1 recognition sequence and serve as generalized transcriptional repressors. *Mol. Cell. Biochem.* 178: 187-196.

Gauthier, Mona L., and O'Day+ H. O'Day, 2001. Detection of Calmodulin-Binding Proteins and Calmodulin-Dependent Phosphorylation Linked to Calmodulin-Dependent Chemotaxis to Folic and cAMP in *Dictyostelium. Cell. Signal.* 13: 575-584.

Giampuzzi, M., R. Oleggini and A. Di Donato, 2003. Demonstration of in vitro interaction between tumor suppressor lysyl oxidase and histones H1 and H2: definition of regions involved. *Biochim. Biophys. Acta* 1648: 245-251.

Grasser, F.A., C. Sauder, P. Haiss, A. Hille, S. Konig, S. Gottel, E. Kremmer, H.P. Leinenbach, M. Zeppezauer and N. Mueller-Lantzsch, 1993. Immunological detection of proteins associated with the Epstein-Barr virus nuclear antigen 2A. *Virology* 195: 550-580.

Hammer, J. A., M. Lydan and G. Jung, 1996. *Dictyostelium* MyoJ associates with membranes of the contractile vacuole complex, *Mol. Biol. Cell* 7: 374.

Hauser, L. Dhar, M. and D. Olins, 1995. *Dictyostelium discoideum* contains a single-copy gene encoding a unique subtype of histone H1. *Gene* 154: 119-122.

Hengstschläger, M., Knofler, M., Mullner, E. W., Ogris, E., Wintersberger, E. and Wawra, E., 1994. Different regulation of thymidine kinase during the cell cycle of normal versus DNA tumor virus-transformed cells. *J. Biol. Chem.* 269: 13836-13842

Horton, P., and K. Nakai, 1997. Better prediction of protein cellular localization sites with the k nearest neighbors classifier. Proc. Int. Conf. Intell. Syst. Mol. Biol. 5: 147-52.

Kim, J-W. and C.V. Dang, 2005. Multifaceted roles of glycolytic enzymes. *Trends Biochem. Sci.* 30: 142-150.

Knecht, D. A., and W. F. Loomis, 1987. Antisense RNA inactivation of myosin heavy chain gene expression in *Dictyostelium discoideum*, *Science* 236: 1081 – 1085.

Koonin, E. V., S. F. Altschul and P. Bork, 1996. BRCA1 protein products...Functional motifs. *Nat. Genet.* 13: 266 – 268.

Larochelle, D. A., K. K. Vithalani and A. De Lozanne, 1996. A novel member of the *rho* family of small GTP-binding proteins is specifically required for cytokinesis. *J. Cell. Biol.* 133: 1321–1329.

Lee, S., R. Escalante and R. A. Firtel, 1997. A Ras GAP is essential for cytokinesis and spatial patterning in *Dictyostelium*, *Development* 124: 983 - 996.

Lydan, M. A. and D. A. Cotter, 1995. The role of Ca^{2+} during spore germination in *Dictyostelium*: autoactivation is mediated by the mobilization of Ca^{2+} while amoebal emergence requires entry of external Ca^{2+}. *J. Cell Sci.* 108: 1921 - 1930.

Lydan, M.A., and D.H. O'Day, 1988a. Intracellular calcium regulates cell fusion during early sexual development in *Dictyostelium discoideum*: Studies using A23187, CTC, La^{3+}, IP_3 and TMB-8. *J. Cell Sci.* 90: 465-473.

Lydan, M.A., and D.H. O'Day, 1988b. Different developmental functions for calmodulin in *Dictyostelium discoideum*: Trifluoperazine and R24571 both inhibit cell and pronuclear fusion but enhance gamete formation. *Exp. Cell Res.* 178: 51-63.

Lydan, M.A., and D.H. O'Day, 1989. The autoinhibitor of cell and pronuclear fusion in *Dictyostelium discoideum* is a specific inhibitor of calmodulin. *Biochem. Biophys. Res. Comm.*, 164: 1176-1181.

Lydan, M.A. and D.H. O'Day, 1993. Calmodulin and calmodulin-binding proteins in *Dictyostelium discoideum*: Developmental regulation by calcium ions. *Exp. Cell Res.* 205: 134-141.

Lydan, M.A. and D.H. O'Day, 1993. Calmodulin-dependent protein phosphorylation and dephosphorylation during fertilization in *Dictyostelium discoideum. Biochem. Biophys. Res. Commun.* 192: 1073-1078.

Lydan, M.A., D.A. Cotter and D.H. O'Day, 1994a. Stage-specific changes in protein phosphorylation during spore germination in *Dictyostelium*: Role of calmodulin. *Biochem. Biophys. Res. Commun.* 201: 430-435.

Lydan, M.A., D.A. Cotter and D.H. O'Day, 1994b. Calmodulin function and calmodulin-binding proteins during autoactivation and spore germination in *Dictyostelium discoideum. Cell. Signal.* 6: 751-762.

Maeda, Y., 1997. Cellular and molecular mechanisms of the transition from growth to differentiation in *Dictyostelium* cells, pp. 207-218 In: *Dictyostelium–A Model System for Cell and Developmental Biology*, Universal Academy Press, Tokyo, Japan.

Moerman, A.M., and C. Klein, 1998. *Dictyostelium discoideum* Hsp32 is a resident nucleolar heat-shock protein. *Chromosoma*, 107: 145-154.

Mohanty, S., S. Lee, N. Yadava, M. J. Dealy, R. J. Johnson and R. A. Firtel, 2001. Regulated protein degradation controls PKA function and cell-type differentiation in *Dictyostelium*. *Genes.Dev.* 15: 1435 – 1448.

Myre, M., 2005. Characterization of nucleomorphin, a novel nuclear breast cancer carboxy-terminal-domain containing calmodulin-binding protein from *Dictyostelium discoideum. PhD Dissertation,* Department of Zoology, University of Toronto, Canada.

Myre, M. A., and D. H. O'Day, 2002. Nucleomorphin: A novel, acidic, nuclear calmodulin-binding protein from *Dictyostelium* that regulates nuclear number. *J. Biol. Chem.* 277: 19735-19744.

Myre, M. A., and D. H. O'Day, 2004a. Calmodulin binds to and inhibits the activity of phosphoglycerate kinase. *Biochimica Biophysica Acta, Molecular Cell Research* 1693: 177-183.

Myre, M. A., and D. H. O'Day, 2004b. *Dictyostelium* calcium-binding protein 4a interacts with Nucleomorphin, a BRCT-domain protein that regulates nuclear number. *Biochemical Biophysical Research Communications* 322: 665-671.

Myre, M.A., and D.H. O'Day, 2004c. *Dictyostelium* nucleomorphin is a member of the BRCT-domain family of cell cycle checkpoint proteins. *Biochimica Biophysica Acta, General Subjects*.1675: 192-197.

Myre, M.A., and D.H. O'Day, 2005. An N-terminal nuclear localization sequence but not the calmodulin-binding domain mediates nuclear localization of nucleomorphin, a protein that regulates nuclear number in *Dictyostelium. Biochemical Biophysical Research Communications* 332: 157-166.

Natsukari, N., S. Zhang, R.Nichols and B. Weiss, 1995. Immunocytochemical localization of calmodulin in PC12 cells and its possible interaction with histones. *Neurochem. Int.* 26: 465-476.

O'Day, D.H., 2003. CaMBOT: profiling and characterizing calmodulin-binding proteins, *Cell Signal.* 15: 347-354.

O'Day, D.H., and M.A. Myre, 2004. Is histone H1 a true calmodulin binding protein? *Abcam Chromatin Structure and Function Meeting*, Cancun, Mexico, 16-19 November 2004.

O'Day, D.H., K.E. Lewis and M.A. Lydan, 1995. Amoeboid gametes and fertilization in *Dictyostelium*: Gamete and pronuclear fusion are mediated by calmodulin and its binding proteins. *Memoires du Museum National d'Histoire Naturelle*, Paris 166: 23-36.

O'Day, D.H., J. Rivera and D.R. McConachie, 1987. Appearance and developmental kinetics of a unique cell type in *Dictyostelium discoideum*: Is it the gamete phase of sexual development? *J. Exp. Zool.* 242: 153-159.

O'Day, D.H, M. A. Myre, S. Wagler and M Chatterjee-Chakraborty, 2005a. Isolation and characterization of *Dictyostelium* thymidine kinase 1 as a calmodulin binding protein. *Biochemical Biophysical Research Communications* 331: 1494-1502.

Onions, J., S.Hermann and T.Grundstrom, 2000. A novel type of calmodulin interaction in the inhibition of basic helix-loop-helix transcription factors. *Biochemistry* 39: 4366-4374

Pickart, C. M., 2001.Mechanisms underlying ubiquitination, *Annu. Rev. Biochem.* 70: 503-533.

Polesky, A.J.M. and D.H. O'Day, 1995. Calcineurin is involved in the chemotactic response to cAMP but not folate in *Dictyostelium discoideum. Mol. Biol. Cell* 6: 91a.

Popanda, O., G. Fox and H.W. Thielmann, 1998. Modulation of DNA polymerases α, δ and ε by lactate dehydrogenase and 3-phosphoglycerate kinase, *Biochim. Biophys. Acta.* 1397: 102-117.

Rasmussen, C., and C. Garen, 1993. Activation of Calmodulin-dependent Enzymes Can Be Selectively Inhibited by Histone H1. *J. Biol. Chem.* 268:23788-23791.

Sakamoto, H., K. Nishio, M. Tomisako, H. Kuwayama, Y. Tanaka, I. Suetake, S. Tajima, S. Ogihara, B. Coukell and M. Maeda, 2003. Identification and characterization of novel calcium-binding proteins of *Dictyostelium* and their spatial expression patterns during development, *Dev. Growth Differ.* 45: 507-14.

Saveanu, L., D. Fruci and P.M. van Endert, 2002. Beyond the proteasome: trimming, degradation and generation of MHC class I ligands by auxiliary proteases. *Mol. Immunol.* 39: 203–215.

Siddhartha, E., D.T. Brown, Z.H. Lu, G.H. Leno, S.E. Wellman and D.B. Sittman, 2002. Histone H1 variants differentially inhibit DNA replication through an affinity for chromatin mediated by their carboxyl-terminal domains. *Gene* 292: 173-181.

Sudol, M., K. Sliwa and T. Russo, 2001. Functions of WW domains in the nucleus. *FEBS Lett.* 490: 190-195.

Szabo, S.P., D.H. O'Day and A.H. Chagla, 1982. Cell fusion, nuclear fusion and giant cell differentiation during sexual development of *Dictyostelium discoideum*. *Develop. Biol.* 90: 375-382.

Szabo, S.P., and D.H. O'Day, 1984. The low molecular weight autoinhibitor of sexual development in *Dictyostelium discoideum* inhibits both cell fusion and zygote differentiation. *Can. J. Biochem. Cell Biol.* 62: 722-731.

Tan, J.L., and J.A. Spudich, 1990. *Dictyostelium* myosin light chain kinase. Purification and characterization. *J. Biol. Chem.* 265:13818-24.

Thomas, J.O., 1998. Histone H1: location and role. *Curr. Opin. Cell Biol.* 11: 312-317.

H. Van Der Wel, H.R. Morris, M. Panico, T. Paxton, A. Dell, L. Kaplan, C.M. West, 2002. Molecular cloning and expression of a UDP-N-acetylglucosamine (GlcNAc):hydroxyproline polypeptide GlcNAc-transferase that modifies Skp1 in the cytoplasm of *Dictyostelium*. *J. Biol. Chem.* 277: 46328-46337.

Vazquez-Padua, M.A., C. Risueno and P.H. Fischer, 1989. Regulation of the activation of fluorodeoxyuridine by substrate competition and feedback inhibition in 647V cells. *Cancer Res.* 49: 618-624.

Watson, H.C., and J.A. Littlechild, 1990. Isoenzymes of phosphoglycerate kinase: evolutionary conservation of the structure of this glycolytic enzyme, *Biochem. Soc. Trans.* 18: 187-190.

Weissenmayer, B., K.Boelckeler, A.Lahrz and R.Mutzel, 2005. The calcineurin inhibitor gossypol impairs growth, cell signalling and development in *Dictyostelium discoideum*. *FEMS Microbiol Lett* 242: 19-25.

West, C. M., H. van der Wel and E. A. Gaucher, 2001. Complex glycosylation of Skp1 in *Dictyostelium*: implications for the modification of other eukaryotic cytoplasmic and nuclear proteins. *Glycobiology*, 12: 17R-27R.

Williams, J. G., A. A. Noegel and L. Eichinger, 2005. Manifestations of multicellularity - Dictyostelium reports. Trends Genet. 21: in press.Yamanaka, S., E. Katayama, K-I. Yoshioka, S. Nagaki, M. Yoshida and H. Teraoka, 2002. Nucleosome linker proteins HMGB1 and histone H1 differentially enhance DNA ligation reactions. *Biochem. Bioophys. Res. Commun.* 292: 268-273.

Yap, K. L., J. Kim, K. Truong, M. Sherman, T. Yuan and M. Ikura, 2000. Calmodulin Target Database, *J. Struct. Funct. Genomics* 1: 8-14.

Yoshida, A., 1996. Hematologically important mutations: Molecular abnormalities of phosphoglycerate kinase, *Blood Cells Mol. Disease* 22: 265-267.

Yoshida, A., 1975. Human phosphoglycerate kinase, *Meth. Enzymol.* 42: 144-148.

Zhukovskaya, N.V., M. Fukuzawa, M. Tsujioka, K.A. Jermyn, T. Kawata, T. Abe, M. Zvelebil and J.G. Williams, 2003. Dd-STATb, a *Dictyostelium* STAT protein with a highly aberrant SH2 domain, functions as a regulator of gene expression during growth and early development, *Development* 131: 447- 458.

Zielinski, R. E., 1998. Calmodulin and calmodulin-binding proteins in plants. Annual Rev. Plant Physiol. *Plant Mol. Biol.* 49: 697-725.

Zlatanova, J. Caiafa, P. and K. van Holde, 2000. Linker histone binding and displacement: versatile mechanism for transcriptional regulation. *FASEB J.* 14:1697-1704.

Index

A

B

C

D

E

F

G

H

I

K

L

M

N

O

P

Q

R

S

T

U

V

W

X

Y

Z